KB043064

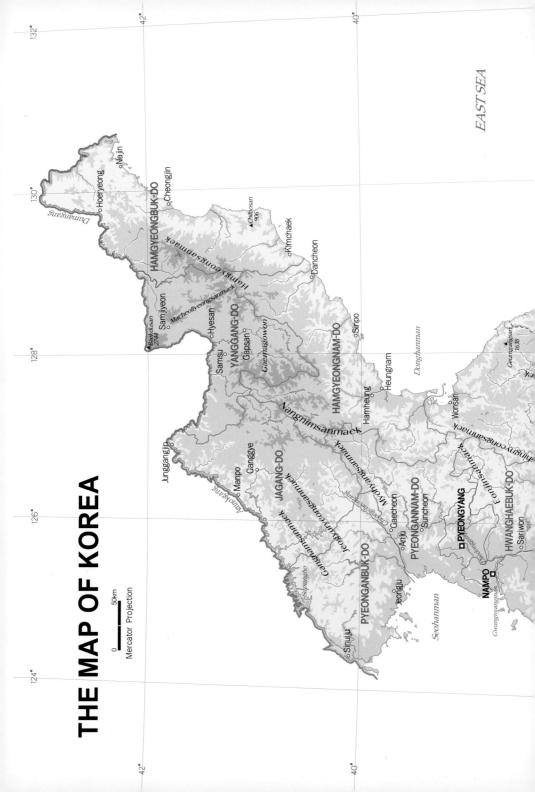

THE MAP OF KOREA

0 50km

Mercator Projection

EAST SEA

HAMGYEONGBUK-DO

YANGGANG-DO

HAMGYEONGNAM-DO

JAGANG-DO

PYEONGANBUK-DO

PYEONGANNAM-DO

HWANGHAEBUK-DO

□ **PYEONGYANG**

NAMPO □

Hoeryeong

Najin

Cheongjin

Kimchaek

Dancheon

Sinpo

▲Chilbosan
906

Baekdusan
2744

Samjiyeon

Hyesan

Samsu

Gapsan

Gaemagowon

Heungnam

Hamheung

Wonsan

▲Geumgangsan
1638

Junggangjin

Manpo

Ganggye

Gaecheon

Anju

Suncheon

Jeongju

Sinuiju

Sariwon

Hambyeongsanmaek

Macheollyeongsanmaek

Nangnimsanmaek

Myohyangsanmaek

Eonjinsanmaek

Cheongnyeongsanmaek

Jeoksyeongsanmaek

Gangnamsanmaek

Dumangang

Amnokgang

Cheongcheongang

Daedonggang

Donghanman

Seohanman

Gwangnyangman

Sipungho

132°

130°

128°

126°

124°

42°

40°

42°

40°

The Climate and Culture of Korea

The Climate and Culture of Korea

by Seungho Lee

PURENGIL

Preface

Today, climate is one of the common interests all around the world. The impact of climate is sometimes much too heavily emphasized or even exaggerated by the news media. It is often the case that all environmental issues are reported as the results of the climate change, which is not true. Desertification is, for example, not always caused by the climate change, but the news media report it as if it is mostly the case. In many cases, desertification is resulted from inappropriate water management experiences by human being. Ordinarily people are much too serious about the climate change than what it may be.

This book does not deal with the problem of climatic change per se. The importance of climate is emphasized only in the first part of the book. That is, the book simply explains why we should understand the climate. Except that, this book pursues to tell the nature and culture of Korea. It is written for Korean, however, there are included rather difficult parts for foreign readers to understand. We could not avoid to face this problem because we are faced in different climatic environment and culture. In the next edition, if available, we would like to try a better way to overcome this problem.

This book largely consists of four chapters. First, the book explains climatic factors which make the Korea's climate. Second, it tells about season, important to understand Korea. Every country has seasons but for Korea which is located at the eastern edge of the Eurasian continent, it is more significant. Particularly Korea has four distinct seasons with a rainy season called *Jangma*. Third, it explains climatic elements of Korea such as temperature, precipitation, wind, fog and frost. Finally, the book explains how climate effects on Korean's life. Wind, snow, heat wave and cold are particularly focused among the climate elements as they are regarded significantly affecting Korean's life.

The book provides several pictures to tell climate and landscape. The author's own experience related to climate is also addressed. Most of the pictures are the author's, while some from acquaintances'. The author especially thanks Korea University's honorary professor, Dr. Kwon Hyuk Jae and Ms. Park Chang Yeon in Chuncheon for their invaluable help. The pictures in the book are not inserted to show the beauty of Korea but Korea's climatic phenomenon, which explains why the book has some ugly pictures.

The author would be very happy if this book helps foreign readers understand Korea better. The author would like to thank Jinna Park for her translation work that enables this book published. The author also thank Korea Foundation for their funding that enables this book translated.

June 2010

Seungho Lee

CONTENTS

Contents ▸▸▸

Contents ▸▸▸

01

Why do we need to know about the climate?

With industrial development and increased consumption, our dependency on the climate is gradually growing. The reason we are interested in climate is the massive influence it exerts on our everyday life. In addition, as global warming continues, climate issues affect our very survival. Today, people take an interest in the climate not only with regard to their own financial futures, but for a graver concern: the future existence of our planet.

oday one has to know about the climate in order to take part in casual discussions of world affairs. This by itself indicates how much the subject has started to concern ordinary people. Articles and news reports about the climate appear frequently in the newspapers or on television. Climate has become a very familiar topic in our lives. What has brought it so urgently to our attention?

The main factor is, above all, the global warming that is occurring throughout the world. From the advent of climate observation with scientific instruments in the late 19th century, the temperature repeated the cycle of rising and falling until the late 1970s. However, from the 1980s, the temperature has steadily increased in all parts of the world. As a result, the past ten years was the hottest decade on record, with 1998 recording the Earth's highest average temperature, and the years 2003 and 2005, the second highest.

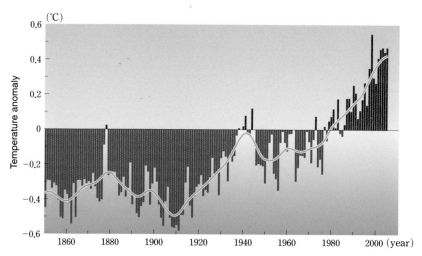

Changes in the worldwide mean surface-air temperatures From the advent of modern meteorological observation, the temperature repeated the cycle of rising and falling, but from the 1980s, the temperature is steadily rising in all parts of the world. This graph represents the anomaly in the average absolute temperatures for the base period 1961 to 1990. (Source: East Anglia University's Climate Research Unit)

The disappearing Rhone glacier As global warming proceeds, there are reports of the glacier melting in the Alps. If one looks at old photographs taken at the Rhone glacier in the 19th century, the area is mostly covered with ice. (Rhone Valley, Switzerland. August, 1996)

During the past decade, there has been extensive media coverage of climate change. In the summer of 2002, an extraordinarily severe flood hit central Europe, claiming more than one hundred lives. Earlier, in the spring of that year, there were reports of a new lake created in Italy as the glacier of the Alps melted. The winter of 2001-2002 was recorded as the warmest winter in the United States since meteorological statistics were first recorded.

Many meteorological disasters have occurred recently also in Korea. In the summer of 2002, some 900 mm of rain fell in one day in the area of Gangneung (in Gangwon-do) due to the influence of Typhoon Rusa. As a result, reservoirs and roads collapsed in the area east of the Taebaek Mountains (Taebaeksanmaek), causing massive loss of property and lives. In September 2007, Typhoon Nari hit Jejudo Island, an area whose

Damages from torrential rains in mountain areas In the past, each summer, the areas in the lower reach of rivers mainly suffered inundations, but from the late 1990s, torrential rains also fall in the mountain areas, which cause flood damages. (Pyeongchang, Gangwon-do. August, 2006)

residents were previously unconcerned with any possibility of inundation. In March 2004, a heavy snowstorm paralyzed road travel nationwide, and in December of the following year, a snowstorm hit the west coast with massive damage: destroying houses, livestock farms, and greenhouses. Moreover, torrential rains fall in mountain areas at unexpected times of the year, with considerable loss of property and lives.

These meteorological disasters differ from those of the past. For instance, typhoons in Korea used to be concentrated along the southeast coast, and snowstorms most frequently occurred on Ulleungdo Island or in the mountain area east of the Taebaek Mountains. Heavy rain damage was considered unique to areas downstream of big rivers. Nowadays, however, the varied meteorological disasters are no longer confined to specific areas.

Therefore, climate and weather are coming to concern us as matters of survival. People become naturally interested in the subject not simply to make a little more profit by selling one more product, but as a graver problem: humanity's survival on Earth.

Climate information is useful

We take an interest in the climate because it greatly affects our lives. With industrial development and increased consumption, our dependency on the climate is gradually growing. Work that in the past could be performed without any climatic data requires that information today. By making good use of climatic data, one can make a bigger profit.

For example, climatic data can be crucial to a merchant dealing in winter products such as electric heaters or warm clothes. If a store orders a large quantity of winter electrical goods in expectation of a cold winter, it will suffer a big loss if the prediction proves wrong and the winter is warm. Similarly, a ski resort that sprayed artificial snow on its slopes would suffer economic loss if it snowed that day.

Likewise in the summer, climate information is important when buying and selling food, ice cream and air conditioners. One summer when the heat was setting new records every day, one electrical appliance store that had made good use of weather predictions by stocking a large quantity of air-conditioners made a huge profit. Naturally, those stores who did not pay attention to such information must have regretted their shortsightedness. Furthermore, some companies refer to seasonal weather information in their advertisements. A phrase like "the hottest summer in a hundred years," makes use of that year's climatic data to promote a special sale of home appliances.

At present, climate information is more than simple data; it is an important asset for businessmen. Reflecting this tendency, several private weather forecast companies have appeared. Each of them provides a variety of climatic data and daily forecasts to their clients. Even the Korea Meteorological Administration has gone far beyond the scope of presenting mere weather forecasts, as it did in the past. It now offers diverse, industry-related monthly and weekly weather data services, and also develops and announces various indices relating weather to daily life. Recently, it is providing daily weather forecast service to mobile phone users. The National Emergency Management Agency also provides weather and disaster-related news via mobile phone.

Climate information is also important for increased productivity in the agricultural sector. Today, most farms are run scientifically, and are equipped with weather forecast equipment. One orchard I came upon

The application of the climate information in orchard management When there is a forecast of frost, a climate-disaster prevention system is immediately run; Water is sprinkled from a black hose suspended along stainless poles neatly fixed in regular distance apart along the orchard to form a mist and to prevent frost from falling on the blossoms or the products. (Naju, Jeollanam-do. Feb., 2008)

during one of my field survey was monitoring weather data every hour using an automated meteorological system, which transmits forecasts of relevant weather phenomena, such as frost, to each work division of the orchard and to the regional agricultural technology center. The related workers or organizations pass that information to the farmers of other nearby orchards via mobile-phone messages, and run a climate disaster prevention system. The Korea Meteorological Administration, of course, provides separate agricultural weather information. In addition, other agricultural organizations provide various weather information and climate data.

Weather forecasts are essential even in preparing outdoor events, both large and small. In most cases, in the past, events were held without making use of weather information. Now even street fortunetellers and card players are influenced by the weather. On a cloudy day, naturally people do not gather. The world is now a place where one has to know the weather and the climate for even the most basic daily activities.

What if temperatures continue to rise?

Until the early 1980s, few people paid attention to the climate. Back then one of the biggest reasons I chose to study climate was because there were very few researchers dedicated to the subject in Korea. While I was doing my military service after completing my master's course, I heard a surprising news report over the radio on a bus going towards the base. It was about global warming. That may have been the first climate-related news broadcast I heard. That was in 1986. Finally climate had become a subject of general public interest. The effects of global warming were gradually beginning to be felt in Korea.

Without doubt, the Earth was becoming warmer. Although there are considerable regional variations, the mean temperature throughout the world has risen about 1°C since modern meteorological observation was established. In Korea's case, it has risen slightly more. What will happen if the warming trend continues? The rising temperature by itself will most likely pose grave problems to mankind. But the dire consequences go far beyond mere temperature increases. Consider the sea level.

Global warming is proceeding more quickly in the polar regions than the areas closer to the equator. The former are covered with ice, but with global warming, the glacier and sea ice has started to melt. For many years now, the media have reported reductions in the Arctic and Antarctic. Melted glacier flows into the sea, causing the sea level to rise. Consequently, land near the sea will face an unimaginable catastrophe. Glacier is also melting and disappearing in high mountain areas, such as the Alps. The glacier melted from mountains also contributes to the rising sea level.

Therefore, even small amounts of rain can cause disastrous floods in highly populated low-lying coastal areas. Such threatened coastal regions are found throughout the world, and especially in Southeast Asia. Settlements on a delta or along the lower reaches of a river frequently suffer inundation, and often face damages from the sudden, catastrophic rise in seawater associated with storm surge. The overall rise in sea level resulting from global warming increases the danger to those areas.

Bangladesh lies in the plain of the Ganges and Brahmaputra rivers, with an average altitude of less than 5 meters above sea level. Already a frequent victim of immense flood damages, Bangladesh is expected to be one of the countries most immediately and seriously affected by global warming. According to climate experts, sea level will rise nearly 2 meters above its current level by 2100. That would entail the submersion of some twenty percent of Bangladesh. Other Southeast Asian nations, including Vietnam, Cambodia, and Burma, are in the same situation, facing the same fate.

The rising sea level will not only reduce the habitable land of an area; it can also threaten the very existence of an entire country. Already many

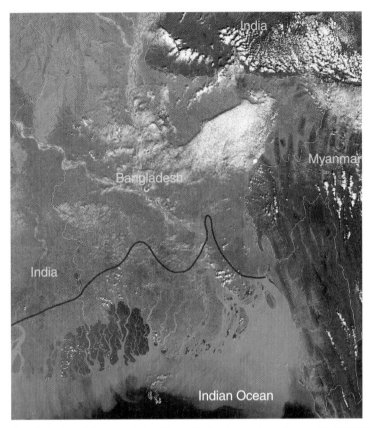

Bangladesh and its vast delta The delta developed along the mouth of the Ganges and Brahmaputra rivers make up a considerable part of Bangladesh territory. The red line indicates the 2 meter above sea level. (Satellite image. Source: NASA)

years ago we heard the news of the world's first "climate refugees" – from a coral island in the South Pacific, called Tuvalu, which is disappearing into the sea.

Global warming is also affecting the ecosystem. Even in Korea, changed features are noticeable in plants that are sensitive to climate. For example, bamboo trees that used to grow only south of the Charyeong Mountains

Bamboo trees growing in the environs of Seoul Bamboo trees known to grow in the warmer regions, south of Charyeong Mountains, are now commonly found in Seoul and its environs. In the photo, on the left is Hangang river and Mt. Ungilsan in the distance. (Yangpyeong, Gyeonggi-do. March, 2008)

(Charyeongsanmaek) are now commonly found in Seoul and its environs. A thick bamboo forest has been firmly established for some time near Asan and Taean (in Chungcheongnam-do). There are also reports that in high mountain areas plants peculiar to the mountains are facing the danger of extinction. It is certain that now Koreans should also be more concerned about climate change, and should be taking some sort of countermeasures.

Why has extreme weather become more common recently?

As the Earth became warmer, extreme weather, such as torrential rains, snowstorms, storms, and droughts, has become more frequent. Regions that used to have little rainfall now often suffer from flooding after a sudden precipitation. And there are snowstorms that isolate areas or consecutive days of heat waves at places where snowfall used to be rare.

Moreover, prolonged droughts cause massive damages to farming.

Korea is experiencing more instances of heavy rain and snow than in the past. It is now quite common for several hundred millimeters of rain to fall in one day. In some regions, torrential rain hits the same area several times a year, worsening the flood damage there. There are also sudden, unexpected snowfalls causing severe damage. In the past, such phenomena were rare occurrences; now, however, they are happening repeatedly each year.

These phenomena are referred to as 'abnormal weather'. Why, then, has abnormal weather become so frequent in recent years? Most climate experts attribute it to global warming. But if such extreme meteorological events continue on a frequent basis, the day will soon come when they cease to be considered abnormal.

Along with the rising global temperature, there are marked changes in

Damages caused by heavy snow Sudden snowstorms isolate villages and cause greenhouses to collapse. (Buan, Jeollabuk-do. Jan. 2006)

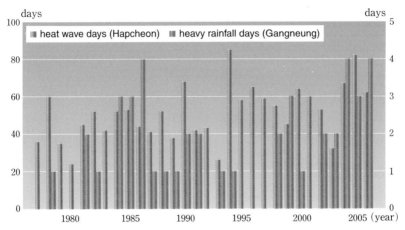

The changes in the number of days of heat waves and heavy rains Recently, as global warming proceeds, the frequency of abnormal weather is sharply increasing.

climate patterns throughout the world. Regional differences in the rate of temperature increase alter the pattern of General circulation, which is responsible for the global distribution of climatic conditions. The possible changes are beyond imagination. Before long, school children may be studying a completely different world map of climatic regions.

As we've seen on several occasions that the whole world has to be alert to possible local effects stemming from a 1°C rise in the temperature of the Pacific Ocean near the equator. Each region of the world responds differently during an El Niño and the effects are extremely difficult to predict. As far as we live from the equator, we are living in an era when the Korean Meteorological Administration concerns itself with forecasts of an El Niño in the waters of the equatorial Pacific. Research studies have indicated many significant El Niño effects on weather of Korea.

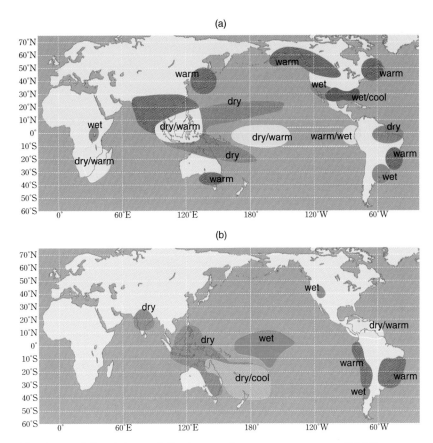

The characteristics of abnormal climate per region upon an occurrence of El Niño
(a) indicates winter, and (b), summer. (Source: NOAA (National Oceanic and Atmospheric Administration) / CPC (Climate Prediction Center))

Efforts to solve climatic problems

According to climate researchers, global warming is the result of human activity in the modern industrial era. In particular, the burning of fossil fuels emits gases, mainly carbon dioxide, into the atmosphere, where they produce a 'greenhouse' effect. In the name of economic progress and

A film about extreme-climate disasters
As climate change becomes a serious social problem, disaster films featuring the subject are also becoming popular.

improved welfare, industrialized mankind has been digging its own grave.

Today, governments and researchers are working to reduce global warming, and to adapt to the climate changes. They have gone so far as to run a market where countries buy and sell carbon dioxide emission allowances, and climate change has become an important agenda at the United Nations.

At present, the climate is changing at an alarming pace. So there is an urgent need to tackle the associated problems. Neglect of those problems would likely lead to serious natural disasters. From the end of the late 1990s, movies have been made about climate-related catastrophes. Many of these have been box-office hits. Concerned and fearful for the future of our planet, more and more people are paying attention to climate change.

By their very nature, climate problems cannot be understood or dealt with within the context of just one country. Most problems transcend national boundaries. Therefore, there are various organizations active under the UN to address climate issues. A representative example is the Intergovernmental Panel on Climate Change (IPCC), whose participants include Korean government representatives and researchers. The IPCC raises awareness of climate change by publishing reports. Their fourth report was released in 2007.

In Korea many researchers are also working on climate problems. These include the recently established Korean Panel on Climate Change (KPCC), whose active members include climate experts from both government and nongovernmental research organizations.

But the efforts of the government and academic institutions are not sufficient to deal with the wide range of complex problems stemming from climate change. More than anything, the interest and effort of ordinary citizens is required. Yet it is only quite recently that most people have become used to hearing the word 'climate'. Now, although many people use the word, its meaning is too often misunderstood.

Part 1

Factors of
the Korean Climate

02

The Korean peninsula is in the middle latitudes on the eastern edge of the Eurasian continent

The geographical location of Korea can be thought of in several different aspects. In terms of climate, the fact that Korea is located in the middle latitudes is most important. By virtue of its location in the temperate zone, Korea has four distinct seasons. Furthermore, lying on the eastern edge of the enormous continent of Eurasia, Korea has more sharply defined seasonal changes than countries located further west on the continent.

A ll our lives, we are constantly fighting for seats: in a classroom or a movie theater, people do their best to grab a good seat. The first thing one does on boarding a bus or a train is to head straight for a comfortable seat. That is because a thing's value often depends on its position.

The case is similar with weather and climate: their qualities are determined by location. However, unlike your seat on a bus or subway, the geographical location of a country is decided once and for all. It cannot be changed. Therefore, people live in the weather and climate determined by their countries' location, adapting themselves to the meteorological conditions.

Korea is in the middle latitudes The Korean peninsula is in the middle latitudes on the eastern edge of the Eurasian continent.

Why do seasons change?

The geographical location of Korea can be thought of in several different aspects. In terms of climate, the fact that Korea is located in the middle latitudes is most important. Korea's location in the mid-latitude is the key factor in its characteristic pattern of seasonal variations.

The axis around which the Earth revolves is tilted 23.5 degrees. This produces variations in the relative altitude of the sun. Seen from the Earth, the sun moves between the latitudes of 23.5 degrees south and 23.5 degrees north. Those lines are respectively called the Tropic of Capricorn and the Tropic of Cancer. These two lines of latitude represent the extremes of the sun's apparent movement relative to the earth. When those points are reached, we experience a solstice: Winter Solstice at the Tropic of Capricorn, and Summer Solstice at the Tropic of Cancer. The change in relative solar altitude is greater at higher latitudes, and smaller nearer to the equator. The change is most extreme furthest from the equator, that is, at

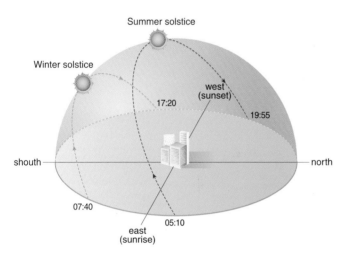

Korea's location and the apparent movement of the sun per season Being in the middle latitudes, there are changes in the solar altitude throughout the year and thereby result in distinctive four seasons.

the polar latitudes. Thus in arctic regions there are days in the summer when the sun is visible for an entire 24 hour period, and days in the winter when 24 hours pass without the sun coming into view at all.

The fact that the Earth is round is an important factor determining the amount of energy that any given region receives from the sun. Though two areas may receive sunlight for the same length of time, more energy is received when the sun is shining from a vertical angle, that is, straight overhead. However, if the sun is shining from an oblique angle, the amount of energy is correspondingly reduced. Therefore, the amount of solar energy received by a given area on the surface of the Earth changes not only according to the season, but also depending on the latitude where it is located.

In short, the heat balance differs per latitude because of the differences in solar altitude, and because the Earth's surface is round. In lower latitudes, closer to the Equator, the surface receives a lot of solar energy,

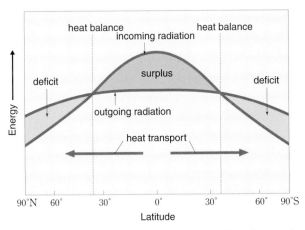

The distribution of energy per latitude The Earth's surface receives an excessive amount of heat the year round in lower latitudes, where the solar altitude is high, while it lacks in heat throughout the year in higher latitudes, where the solar altitude is low. In the middle latitudes, the heat balance differs per season.

while in the high-latitude polar regions, the surface receives very little energy from the sun. The thermal difference between the two regions is phenomenal.

The flow of energy is not one-way: in addition to absorbing solar energy, the Earth's surface also emits energy – predominantly reflected solar energy – into the atmosphere, and out into space. But the difference between the amount of energy emitted from the polar regions and the amount emitted near the Equator is minimal compared to the enormous difference in the amounts of energy those regions receive from the sun. Therefore, in equatorial and tropical latitudes, the amount of energy reflected back into the atmosphere is small compared to the amount of solar energy the surface absorbs; the situation is reversed at polar latitudes. As a result, there is always an excessive amount of heat energy in low latitudes, and a relative lack in polar regions. That is the main reason it is unbearably hot in the tropics, and frightfully cold near the poles.

Because Korea lies in the middle latitudes, the annual changes in apparent solar altitude are neither as extreme as at the poles nor as minimal as near the equator. Over the course of a year, there is a rough balance between the amount of energy received from the sun and the amount reflected and emited from the Earth's surface. However, during the season when the solar altitude is high, we experience a hot summer as the surface receives more solar energy, and when the solar altitude is low, we have a cold winter as less heat is received from the sun. And between those two seasons, there are autumn, and spring. During the latter two periods, there is neither too much nor too little heat, so the weather is pleasantly cool or warm.

The pattern of heat distribution at different latitudes can be likened to a house in winter with a heated stove inside. Cold air surrounds the house, but indoors it is warmed by the heat from the stove. The air outside the house is not being heated, so it remains cold. The window pane that forms a boundary between those warm and cold areas is analogous to the

In the streets of western Europe in summer In Korea it's midsummer, but in Ireland, it's cold enough to wear a thick overcoat. However, the some Irishmen are wearing short-sleeve shirts. (Galway, Ireland. July, 2004)

temperate region of the middle latitudes.

Korea's eastern location: why it matters

During a temporary sojourn in the west of the Eurasian continent, I assumed I'd easily adapt to the local climate. But in the event, the expectations and habits I'd acquired in Korea left me unprepared for the weather I encountered. It was most awkward when, dressed in a manner appropriate to the current season in Korea, I would notice how different my attire was from what the locals were wearing. In midwinter, they were walking downtown in short-sleeve shirts, something unimaginable in Korea. And several times during their midsummer, I wished I was wearing a thick overcoat. To be sure, they had seasons, but the variations between them differed markedly from what we experience in Korea.

Korea's climate is crucially determined by its location in the mid-latitude zone, and at the eastern edge Eurasia, the world's largest continent. Being

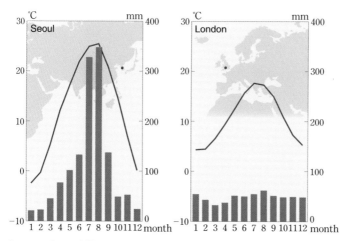

A comparison of Korean and western European climate In Korea, there is a big seasonal variation in temperature and in rainfall, however, in western Europe, the amount of rainfall is even the year round, and the temperature difference between the summer and the winter is not as drastic as that in Korea.

on the eastern edge of Eurasia, Korea is affected by both continental and oceanic influences. The year-round marine influence makes for a climate quite different from countries further to the west, which are far from the ocean.

In Korea, it is cold and dry in the winter, and hot and rainy in the summer. Those are the general characteristics of the East Asian climate, and they reflect the combined influences of the continent and the ocean. When the Korean peninsula is under the northwesterly influence of the continent, the weather is cold and dry; but when the southern ocean exerts its influence, the weather is hot and humid.

The climate of a region decisively affects the local culture. Korea's sweltering summers, and freezing winters are responsible for a distinctive element of its culture: preserved food, such as winter kimchi.

The wind in the mid-latitudes blows from west to east throughout the year. Known at the 'westerlies', these winds are an important force in moving the region's air from the west to the east. All weather appearing in the mid-latitude region is affected by the westerlies. Torrential rain, a long rainy season, a prolonged drought, continuous warm weather in winter, or a long-lasting cold spell – they are all due to the effect of the westerlies. Even the yellow dust that plagues Korea, especially in springtime, is dust carried by the westerlies from the Huangtu Plateau (Loess Plateau) or the desert in China.

The western side of the Eurasian continent is exposed to the vast, open expanse of the Atlantic Ocean, so there is hardly any barrier against the westerlies. In that region, strong winds often last for several days. Such winds are much more powerful than Korea's typhoons, which typically last no more than a day. Consequently, windmills were developed early in western Europe to make use of the abundant wind. Even today, wind power is an important source of energy in western Europe. The force of the westerlies weakens as it crosses over the Eurasian continent and encounters

Windmills of western Europe In western Europe which is strongly affected by the westerlies throughout the year, wind power was developed early on as a source of energy. (Zaanse Schans, the Netherlands. July, 2004).

high mountains. Thus, the direct effect of the westerlies is weaker in Korea than it is along the western coast of the continent. Instead, it is the difference in temperature between the vast Eurasian continent and the Pacific Ocean that accounts for most of the wind in Korea. That is to say, the important factor generating the wind around Korea is the thermal difference between the northern Pacific Ocean and the Siberian plain.

When the Siberian plain cools down and freezes in the winter, the air becomes much more dense there than the air over the northern Pacific Ocean. Since air moves away from areas of higher density towards areas where it is less dense, a strong wind blows from the Siberian plain towards the Northern Pacific Ocean. That is the northwesterly winter monsoon which usually affects Korea in winter. The northwesterly winter monsoon brings the cold and dry Siberian air, so when this wind blows, it is extremely cold and dry.

In the summer, the Siberian plain quickly heats up. But since water

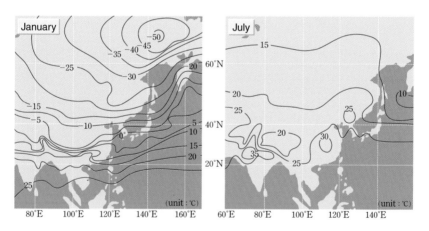

The summer and winter temperature distribution in Korea and in the nearby region The northwesterly winter monsoon brings the cold and dry Siberian air to the peninsula in winter, and the southwesterly summer monsoon brings hot and humid air from the Pacific Ocean in summer.

changes temperature much more slowly than land, the temperature of the Pacific Ocean hardly shows any change. Therefore, the air density above the Pacific Ocean in the summer rises, and the wind blows from there towards the continent. This wind carries the muggy, humid air from above the Northern Pacific, so it is very hot and brings a lot of rain. However, the difference in the air densities of these two is not as great in summer as in winter. Therefore, the wind is not as strong. The wind is fiercer, the greater the temperature range is between two regions.

How have the sharply defined seasons shaped Korean culture?

There was a time when I believed that the hurried pace and constant rushing that characterize life in Korea were due to the capricious Korean weather with its sudden, unpredictable changes. But my experience on the western side of the Eurasian continent convinced me otherwise. The weather of Ireland, an island country off the west coast of the Eurasian continent, was indeed unpredictable beyond words. One could experience the weather of a whole year in a single day. Yet there was nothing hurried about the Irish lifestyle. Instead, compared to Korea, the tempo in Ireland was so slow that I found it hard to adjust.

After that experience, I came to think that the fast pace and rushed quality of Korean life may not be due to the rapid, unpredictable changes in daily weather, but rather to the change of seasons. Since Korea has distinct seasons, there is great significance to each time of year. This was especially true when Korea was a primarily agricultural country. Particularly in the regions where they planted two crops a year, farmers had to make very good use of their precious time to harvest their first crop and then plant their second one. Missing the right moment did not just mean falling behind in the competition with others; it could result in destitution or even starvation. Therefore, one had to hurry through the agricultural cycle, and that habit may have continued into the modern era.

Korea's four seasons The sharp change of seasons in Korea beautifully transforms the scenery of the peninsula. (From top left, clockwise, Yeongdeok, Gyeongsangbuk-do (April, 2006); Hyeopje, Jeju-do (June, 2007), Pideongnyeong, Gangwon-do (Jan., 2008), and Baegyang Temple, Jeollanam-do (Oct., 2006))

The sharp change of seasons in Korea beautifully transforms the scenery of the peninsula: pink, yellow, and white flowers in spring; the blue sea and green trees and grass in the summer; the golden yellow rice fields and red maple leaves in the autumn; and the mountains covered with white snow in the winter. All scenery is precious to us. And the change of seasons adds to the natural scenic beauty of Korea.

The sharply defined seasons have also had a powerful influence on various aspects of daily life. The architecture of traditional Korean houses (*hanok*) reflects the need to survive severe winter cold yet remain comfortable in the muggy summer heat. I remember how one foreign climatologist was surprised to see how a *hanok* was built to accommodate the changes of the weather. He had seen the *hanok* at the Cheongpung Cultural Properties Complex, and had noted that there was a thick wall behind the main wooden-floored hall (*daecheong*) to block the wind during the winter. But in order to provide an open space for air to circulate during the heat of summer, there was no wall or door at the very front of the house. That was the style of most *hanok* built in the central part of the Korean peninsula.

Hanok usually faced south. With that orientation to the sun, and with eaves of the proper length, the roofs absorbed heat from the sun in the winter, but also provided shade in the summer. Similarly, the placement of a large flat stone (*gudeuljang*) as part of the heated flooring system (*ondol*) was designed to keep the rooms cool in the summer and warm in the winter.

In each region of Korea, various seasonal customs developed. Most of them are connected with the climate features of a certain time of the year, such as *jwibullori* on the night before the first full moon of the year by the lunar calendar. *Jwibullori* involves setting fire to the dry hay on rice fields and vegetable beds. The purpose is to burn any insects that have gathered in the hay to avoid the winter cold, and any eggs they have deposited there. At this time of the year there is little risk of the *jwibullori* fire spreading dangerously because the ground is fairly wet from the melting snow. The ashes become fertilizer for the new crops. In the last decade or so, traditional farming customs such as *jwibullori* and *daljip tae'ugi* (the burning of straw straps that were wrapped around trees during the winter) have been turned into local festivals to attract tourists from the big cities.

During my youth in the mountainous area in Jejudo, I saw big fires rising every night in early spring from the direction of Hallasan. I sometimes stayed up all night worrying that the fire might spread down the mountain to my village. Back then farmers on Jeju used to light huge fires called *bang'aetbul* on the fields in late winter and early spring. The fields would be flaming red for several nights. Later the local government

The architecture of traditional Korean houses (*hanok*) *Hanoks* reflect the need to survive severe winter cold (a high wall built facing north) yet remain comfortable in the muggy summer heat (an open wooden floor hall facing south). (Yeongju, Gyeongsangbuk-do. Jan., 2007)

Jwibullori On the night before the first full moon of the year (on the lunar calendar), fire is set to the dry hay on rice fields and vegetable beds to burn any insects that have gathered in the hay to avoid the winter cold. (Museom village, Yeongju, Gyeongsangbuk-do. First full moon of 2008)

outlawed the practice, fearing that it could cause forest fires. However, the farmers needed that fire so much that they devised many tricks to escape the prohibition. They would light a long, home-made fuse and leave their house and village in the morning. By the time the field was ablaze, they were nowhere near it. Because summer is damper on Jejudo Island than elsewhere in Korea, the grass is thicker and coarser. Therefore, in order to raise livestock, farmers on Jeju must burn all the grass left in the fields so that new, soft grass can grow. Today on Jeju this practice is being continued under the name of "Field Fire Festival".

As the traces of these field fires disappear, spring starts on Jeju. During the winter, the cows have been eating *chol*, hay prepared after the autumn harvest by drying the native wild grass. Now, just as the supply of *chol* is nearly depleted, the fields turn green with the growth of fresh spring grass. The cows are released from their pens, free to graze in the fields until early

Field Fire Festival An old practice of farmers burning all the grass left in the field so that new, soft grass can grow is continued togday as an annual festival on Jejudo Island. (Saebyeol Oreum, Jeju. March, 2008. By Lee Jun-taek)

A cattle farm in early spring on Jejudo Island When the supply of dry hay is nearly depleted in early spring, the cows are released from their pens, free to graze in the fields. (Jedong Farm, Jejudo. April, 2008)

autumn.

On the 105th day after the winter solstice, Koreans celebrate the holiday of *hansik*. The literal meaning of *hansik* is 'cold meal', referring to the cold rice which is customarily eaten on that day. And this custom is also closely related to the climate. Because this is the driest period of the year throughout Korea, any carelessness with fire can easily lead to huge disaster. So, the use of cooking fires was forbidden on *hansik*. Hence the customary cold rice. Custom notwithstanding, every year at around this time, there are news reports of isolated forest fires throughout Korea.

Another traditional closely related to the climate is the Water Greeting celebrated on the seventh day of the seventh month of the lunar calendar by drinking or bathing in natural spring water. Tradition attributed especially beneficial medicinal effects to any rain that fell on that day. Bathing in that water was believed to be an effective remedy for dermatological problems. In fact this day comes just after the end of Korea's rainy season, a long period of frequent rains which wash the dust from the air. Naturally enough, this would have been regarded as the time when the water running in streams throughout Korea was at its cleanest.

Since the amount of rainfall in Korea varies greatly according to the season, the threat of water shortage had to be taken seriously. Going back to the Silla Dynasty, in the 3rd century C.E., there are records of Koreans building large reservoirs to maintain the year-round supply of water. As another consequence of this concern, the careful measurement and recording of rainfall started very early in Korea. Indeed, the rain-gauge introduced in mid 15th-centruy Korea is believed to be the first scientific device for quantifying rainfall in the world.

Old reservoirs-Uirimji Known for having been built during the Silla Dynasty, in the 3d century C.E., Uirimji is still used as a reservoir today. Its surrounding area has been developed into a public park. (Jecheon, Chungcheongbuk-do. July, 2006)

03

How the Taebaek Mountains affect Korea's climate and culture

In addition to geographic location, there are other factors that determine a region's climate, such as the topography, altitude, and distance from the ocean. In Korea's case, a mountainous, complicated topography, and the three seas surrounding the peninsula combine to produce complex meteorological effects.

The Southern Sea (Jeju Strait) the calm sea south of Korea, is dotted with many islets. Seen from an airplane on a summer day, these are sometimes more impressive than mere islands. At mid-day, rich white cloud formations appear on one side of each islet, but the other side is clear and sunny: a remarkable phenomenon for such tiny land masses. The spectacle provides confirmation of a basic pattern of nature: clouds form only on the windward side of a mountain.

Now if that is what's happening on a small hill of a tiny islet, imagine how crucial a role is played by a major mountain range like the Taebaek Mountains, which runs down the eastern side of the peninsula. By blocking the moisture-laden winds from the sea, they cause the formation of clouds on the eastern side. As a result, Korea's east coast is known for heavy snowfall, while snow is rare on the west coast. But if the Taebaek Mountains were moved over to the western side of the peninsula, the pattern would be reversed: heavy snow on the west coast and virtually none on the east.

In addition to geographical location, several other factors determine the climate features of a region: topography, altitude, and interaction with nearby seas. Since most of Korea's territory is mountainous, the topography is quite complicated. Furthermore, as a peninsula, Korea has water off of three coasts. These seas are a significant factor in the climatic differences between the east and west sides of Korea.

How mountains affect climate

Often when people see clouds forming along the side of a mountain, they say, "The clouds are unable to go over the mountain." As a rough description, that makes sense. However, a more accurate explanation would be that the air is, in fact, passing over the mountain, but the cloud is disappearing in the process. When moving air encounters a mountain and starts to ascend, it forms clouds. Once the air has passed the peak of the

Clouds formed on the hills Clouds form on the windward side of a mountain, but they disappear as the air descends on the other side of the mountain. (Wolchulsan, Yeongam, Jeollanam-do. Aug., 2007)

mountain, it starts to descend. When the air descends, the clouds disappear. But to the observer, it does indeed appear that the clouds cannot manage to climb over the mountain.

The highest mountains in Korea are concentrated on the northern side of the demilitarized zone. Since travel to North Korea is restricted, few of us get a chance to appreciate the majestic scale of those mountains. Still, although not quite as high, the Taebaek Mountains that extend into South Korea constitute a major mountain range, which powerfully influences the climate. In addition, as a formidable natural barrier to cultural interaction, the Taebaek range has historically formed a border between markedly distinct lifestyles.

Today you can easily cross over the Taebaek Mountains by an expressway or a railway, both connected to major cities. You can even fly by commercial airline. But these are quite recent developments of only some

Forests of bamboo trees in Yeongdong resion Yeongdong resion, east of the Taebaek Mountains in Gangwon-do, there are many bamboo-tree forests because the temperature is warm in winter, just like in the southern provinces of the peninsula. (Yangyang, Gangwon-do. Jan., 2008)

thirty years. Not so long ago, the mountain range was so difficult to traverse, and the climates on either side were so different, that sharply distinct lifestyles developed. Thus, although the areas immediately to the east and to the west of the mountains are both within Gangwon-do, the inhabitants speak different dialects and eat different foods. Even the plants are different. East of the mountains, forests of bamboo trees are ubiquitous; but to the west, bamboo is scarcely seen.

Mountains make up nearly 70 percent of Korean territory. For the area south of the DMZ, the figure is slightly over 60 percent. Almost nowhere in the country can you look across the land without a mountain in sight. These mountains lead to remarkable local variations in the climate. Thus, even is this relatively small country, there are several regions with distinctly different climate patterns.

Summer vacationers to the east coast often face an unexpected difficulty.

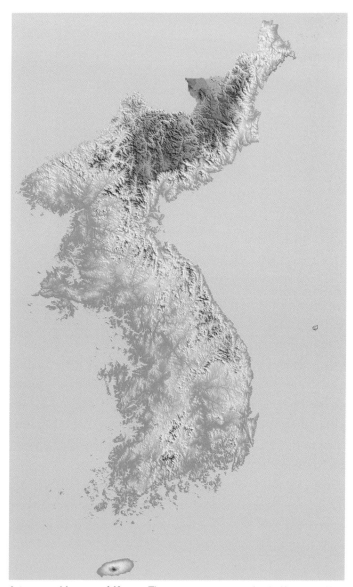

A topographic map of Korea There are many mountains in Korea, making the topography quite complicated, and they are one of the main reasons for regional climate differences.

The summer weather by the Taebaek Mountains When the northeasterly blows, it is grey and gloomy by the east side of the mountain, but on the western side, it is clear and sunny. (Pyeongchang, Gangwon-do. Aug., 2006)

Driving from the west through the Daegwallyeong Pass in the Taebaek Mountains, some 800 meters above sea level on a clear, sunny day, they suddenly find a curtain of dense fog blocking the view ahead.

In fact, that is a cloud clinging to the mountain. Once they are through the Daegwallyeong Pass, the region east of the mountains greets them with a gray sky, and quite often a drizzle. The plan to spend a day on the beach is quickly abandoned; they're actually feeling chilled. After two or three days of the same weather, they give up their holiday plans and cross back through the Pass. No sooner have they crossed the mountains then the rain disappears and they are greeted by a bright sunshine. Such is effect of a mountain range. As the cold, gloomy northeasterly wind blows from the sea, clouds develop on the east side of the Taebaek Mountains. But shielded by the same mountains, the western side is sunny and warm.

There are many other instances where the weather is completely different on either side of the Taebaek Mountains. Torrential rain can be pouring on one side, while the sun beats down on the other. On the same day that saw a record-breaking 900 mm rainfall east of the range in Gangneung, the influence of Typhoon Rusa on the west side of the range brought merely 62.5 mm of rain to Hongcheon, and only 45.5 mm to Seoul. Even when there is a snowstorm east of the mountain range, to the west the skies are usually cloudless and sunny. Koreans are accustomed to hearing news reports of cars stranded by snowstorms on Daegwallyeong Pass when not a flake of snow has fallen on the rest of the country. On Jan. 14, 2003, while a record-breaking 36.8 cm of snow fell on Gangneung, east of the range, it was bright and sunny west of the range, in Hongcheon, Wonju, and Seoul.

When moving air meets a mountain, it ascends the barrier. The act of rising consumes energy from the surrounding air, causing the temperature

Clouds formed on the hills As air ascends a mountain, its temperature falls to dew point, the air condenses as water droplets to form a cloud. People call low clouds on mountains, 'fog'. (Hangyeryeong, Gangwon-do. Aug., 1999)

to drop. When the temperature falls to the dew point, the air condenses as water droplets to form a cloud. When a cloud develops further, the droplets fall as rain or snow. Therefore, it often rains or snows on the windward side of mountains. Mountain trails in Korea have signs posted to warn hikers of fog-prone areas and advise caution. In such places, the 'fog' is actually clouds.

After it yields up its moisture as rain or snow, the air that passes over a mountain is dry. Moreover, it undergoes compression as it descends the mountain. Increased pressure causes the temperature to rise, and when it passes the dew point, the clouds disappear. So, on the lee side of a mountain you encounter few clouds, very little rain, and many clear, sunny days.

That is, in the mountains, there is a lot of rainfall on the side where the wind is blowing, and little rainfall on the other. The reason Gyeongsang buk-do has less precipitation than other regions is the mountains that surround it. No matter which direction the wind is coming from, the air becomes dry while passing over the mountains. This greatly reduces the probability of rain.

Mountain ranges can also act as barriers to the wind. Consequently, in the lee of a mountain, wind velocities are low. Through this effect, the Taebaek Mountains blocks the cold northwesterly winter monsoon. As a result, the eastern side of Korea is not as cold as the western side during winter. The barrier effect from the mountains leads not just to warmer temperatures, but also lower wind velocities than on the western side.

Because mountains break up the pattern of winds, the effect of the chilly northwesterly wind is weakest in areas closest to the mountain range. Therefore, if you are travelling through the eastern, coastal parts of Korea on a day when the northwesterly wind is blowing fiercely, the further you are from the mountains, the colder you'll feel. The bigger the mountain and the closer you are to it, the more warmth you'll enjoy.

The fields of Seogwipo on Jejudo Island in winter Farming in temperate Seogwipo goes on throughout the year. (Seogwipo, Jeju-do. Feb., 2008)

Hallasan, in Jejudo Island, blocks the cold northwesterly wind. Thanks to that mountain barrier, residents of Seogwipo, in the south of the island, enjoy much warmer winters than the citizens of Jeju City, in the north. Indeed, even in the middle of the winter, the weather south of Hallasan feels like early spring. Thus Seogwipo is one of the first places in Korea to announce spring each year. Farming there goes on throughout the year, and subtropical produce is becoming increasingly common.

As altitude rises, temperature drops. According to environmental lapse rate, the air temperature falls by 6.5°C for every 1,000 meters of increased altitude. This explains why it is colder up in the mountains than in the low-lying areas nearby. Thus the high altitude Daegwallyeong Pass has a lower monthly mean temperature than Gangneung, a nearby city at a much lower altitude. Because of the lower temperatures in high altitude regions, the agricultural growing season is shorter. However, farmers can do

Highland Korean cabbage farms On fields at high altitude, Korean cabbages and seed potatoes are grown taking advantage of the cool summer climate. (Anbandeogi, Gangwon-do. Aug., 2007)

well in these areas by taking advantage of the cool summer climate. They can profit by being able to supply farm products that are no longer in season on farms down in the plains.

For instance, the farms on Daegwallyeong Pass made use of the high altitude and the flat land conditions to grow high-value vegetables, such as the Korean cabbages used for kimchi. The high altitude fields are also good for raising sheep and dairy cows. Highland vegetable and livestock farming is practiced in other highland areas as well: in Hoenggye, near Daegwallyeong, and also in Jinan, and in Muju, in Jeollabuk-do.

Korean cabbages and potatoes are the main highland agricultural products. Korean cabbages, however, have become less profitable since kimchi refrigerators have became popular nationwide. Potatoes are still a favorite crop. Particularly profitable are seed potatoes that can be grown on farms at elevations above sea level 700 meters. Although highland

The climate conditions in the mountains Compared to farms down in the plains, farms on high elevation have the advantage of having a continual source of moisture both in the air and on the ground, which helps the growth of plants. (Hallasan, Jeju-do. Aug., 2000)

agriculture can bring good incomes to farmers, the crops require large quantities of pesticides because they are grown in the warm, bug-infested summer season.

Because the temperature is low at high altitudes, the relative humidity is high. Therefore, water vapor easily condenses to form clouds. Rain is frequent and conditions on the ground are generally moist. The clouds or 'fog' formed on the hills is a continual source of moisture. This abundant moisture both in the air and on the ground gives mountainous areas an advantage over lowlands for the growth of plants. However, the frequent rainfall in such areas is an inconvenience, and sometimes even a danger for hikers.

How the seas surrounding Korea affect its weather
If Korea were not geographically bordered by water on three sides, would

the weather predictions be more reliable? Influences from the sea present serious complications to weather forecasters. The sea is subject to changes of its own: gradual changes in water temperature and in the direction of currents. And since even very slight changes in sea temperature produce enormous meteorological effects, weather prediction is never a simple task with a sea nearby.

The complications are especially pronounced for Korea, a peninsula surrounded by three seas: the Yellow Sea, the Southern Sea (Jeju Strait), and the East Sea. Each of these bodies of water has different characteristics, so their influences on the peninsula also differ. Along with the mountainous topography of the peninsula, this diverse marine influence is the major reason for the remarkable differences in the climates of Korea's various regions. Areas close to the sea contrast sharply with inland areas. For example, the regions with heavy snowfall are close to the sea: Ulleungdo Island, the east coast of Gangwon-do, and the west coast of Jeolla-do.

Among Koreans, Ulleungdo Island is famous for prodigious amounts of snow, sometimes reaching the eaves of local houses. The frequent and abundant snowfall results from Ulleung's location as an island on the East Sea, and the island's commanding mountain: Seonginbong Peak, towering 984 meters above sea level. Clouds form when the northwesterly wind carries cold Siberian air across the relatively warm East Sea. When these clouds reach Ulleungdo Island, they come up against Seonginbong Peak. Their rapid ascent of the mountain causes the clouds to thicken, resulting in snowstorms. Residents of the island wrap a layer of straw (*udegi*) around the eaves of their houses to protect their roofs from the snowstorms.

The heavy late-winter or early-spring snowfalls along the east coast in Gangwon-do are also products of the East Sea's influence. Once midwinter is over, the cohesion of the Siberian air mass weakens. In late winter, as it disperses and the center passes the northern part of the Korean peninsula, the northeasterly wind blows. This cold air from the northeast meets with

Houses on Albong, Ulleungdo Island A house is almost completely buried in snow. The prodigious amounts of snow is the reason houses on Ulleung have steep roofs and a layer of straw (*udegi*) around the eaves of their houses to protect their roofs. (Ulleungdo Island, Gyeongsangbuk-do. Jan., 2003)

the warm water of the East Sea and creates clouds, which continue moving until they come up against the Taebaek Mountains. Once again, as the clouds ascend the mountain, they thicken. Precipitation in the form of snow is the final result.

The influence of the sea is also responsible for the heavy snow that hits Jeolla-do on the southwest coast. The cold air blown from Siberia meets with the warm sea water. The temperature difference causes clouds to form. Flying from Seoul to Jeju in winter, if you look out the window, you'll see these clouds spread above the Yellow Sea like a cotton-wool blanket.

The Kuroshio Current is a strong flow of warm, tropical water from off of Taiwan northeastward past the Yellow Sea and East Sea. Like the Gulf

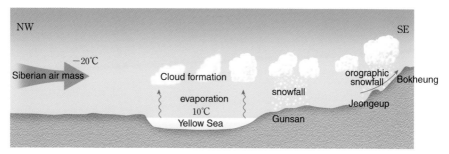

The heavy snow in the Jeolla-do The cold air blown from Siberia meets with the warm air above the sea and the temperature difference causes clouds to form. When these clouds reach the peninsula, they produce snow in places along the west coast, especially in the Jeolla-do.

Stream in the Atlantic Ocean, it brings the warming effects of tropical waters far north towards the arctic. The warmth of this water is in sharp thermal contrast to the cold northwesterly wind blowing down from Siberia. While that wind can bring the air temperature above the Yellow Sea down to -20°C, thanks to the warming Kuroshio Current, the water itself will be around 10°C. Thus, the air closest to the sea becomes much warmer the air higher up, which leads to powerful convection effects. Known as the Sea Effect, this is the reason clouds develop over the vast area of the warmed sea that comes into contact with the cold Siberian air.

When the temperature differential is extreme, cumulus clouds are formed. Although people associate these cumulus with summer weather, they can produce thunder and lightning even in midwinter. When the cumulus reach the land, they produce snow in places along the west coast, such as Gunsan, Yeonggwang, and Buan (in the Jeolla-do). Then, after a temporary respite, they move further inland until they come up against the Noryeong Mountains(Noryeongsanmaek). There, once again, as clouds ascend mountains, the result is snow.

The snow generated by the Sea Effect usually appears along the west coast of Korea, mainly from Taeanbando in Chungcheongnam-do to the

Drifting ice in the lower reaches of the Hangang river In midwinter often there are extensive formations of ice in the lower reaches of the Hangang. (Gimpo, Gyeonggi-do. Jan., 2006)

coastline of Jeollanam-do. The water in Gyeonggiman bay is cooler than the sea further south, so the Sea Effect rarely occurs there. The reason for the relatively low sea surface temperatures in Gyeonggiman is the inflow of cold water from the Hangang and Imjingang. In midwinter there are even extensive formations of ice in the lower reaches of the Hangang.

The warmest current to be found in any of the seas off of Korea flows through the water around Jejudo Island. When the chilly Siberian air mass approaches this warm water, the air becomes extremely unstable. This leads to strong convection, pushing the air currents to higher altitudes, ultimately producing snowflakes of the size and solidity of hailstones. Walking outside during such a snowfall is a difficult and painful experience, as the snow stings your face and body.

When I first left my native Jejudo Island, I lived in an old traditional house on a hill in downtown Seoul. When I opened the window, I had a

Clouds formed by the Siberian air mass in winter The temperature difference between cold air blown from Siberia and warmer air above the Yellow Sea causes thick clouds to form in winter. (Jeju, Jeju-do. Feb., 2008)

bird's-eye view of the tile roofs of other houses. One winter morning, I was surprised to see the paper-paned window shining bright. When I opened it, I saw that all the tile roofs in the neighborhood had turned completely white overnight while I'd slept peacefully. I experienced a moment of culture shock as I recalled nighttime snowfalls on Jeju, where the loud plink of snow hitting the galvanized metal roofs made it difficult to sleep.

04

Air masses: a tale of outside influences

Because of its geographical location, Korea is affected by diverse air masses. In particular, the Korean peninsula is subject to the influence of two air masses from an area much colder, and another from an area much hotter than itself. So, during a given time of year the weather can be completely different depending on which air mass is affecting the country.

A ir is distinguished from other materials by how fast it moves. What's more, as part of its movement, air can circulate back to its original position. Water can also move quite fast, but even at its fastest, it's no match for a powerful flow of air. A region's weather is determined by the characteristics of the air that is passing through its skies.

Located in the middle latitudes, Korea is affected by diverse air masses. The country is affected by two cold air masses from the cold north, and a two others from the hotter region in the south. So during any given period, there can be completely different weather depending on which air mass is affecting the country.

The main air masses affecting Korean weather are developed in three different places: the Siberian plains, the Sea of Okhotsk, and in the vast sea of the North Pacific Ocean. When the air approaching the peninsula is from a cold region, it brings winter weather; air from a hot region brings

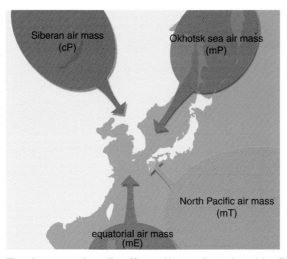

The air masses that affect Korea Korean climate is mainly affected by two cold air masses from the north, and two others from the hotter region in the south. Namely, they are air masses developed in three different places: the Siberian plains, the Sea of Okhotsk, and the vast sea of the Pacific Ocean.

summer weather.

Air from the Siberian plains

Just thinking of the air from the Siberian plains approaching Korea gives me the chills. Though it's nothing compared with the cold of Siberia itself, walking to school in the winter on Jejudo Island felt every bit as chilly. The school was in the northwest of my house, and as I walked through bright open fields early every morning, my nose, fingertips and toes all froze. And if there happened to be a snowstorm, I could easily imagine myself in Siberia. However, my most intense experience of cold didn't come until years later, when I was doing my military service in the valleys of Gangwon-do. Going right through my woolen army uniform to penetrate my body, that cold was incomparably worse than anything I'd suffered as a child on the way to school on Jejudo Island.

It is usually in winter that the air mass grows powerful in the Siberian plains. At that high latitude, days become very short in the winter. So not only is the amount of solar energy reaching the earth's surface greatly reduced, but what little energy is received gets reflected by the snow. The cooling effect at night lowers temperatures even further. In addition to the cold, it is also dry in the Siberian plain because it is far from the sea. When the air of Siberia directly affects Korea, the whole country freezes up. Despite prevailing dry conditions, there can be considerable snowfall along the western coasts of Jeolla-do and Chungcheong-do, and in the island areas.

When the peninsula is directly hit by the Siberian air mass, it is common for high winds to blow, cutting off ferry transportation to small islands. These strong winds result from the big temperature difference between the continent near the peninsula and the Pacific Ocean. At such times, visitors to an island nervously pace about, unable to return home, and the piers are full of fishing boats that cannot go out to sea. What is

The sea on a day the northwesterly is blowing fiercely When the peninsula is directly hit by the Siberian air mass, and a storm alert is issued, fishing boats that were out at sea swarm in, searching for a calm harbor. In this photo, one can see the water is calmer where many boats are docked at the pier than out in the near sea. (Southern coast on Jejudo Island. Jan., 2008)

more, boats that were out at sea swarm in, searching for a calm harbor.

The Siberian air mass affects Korean weather for the longest period in a year. In a typical year it shows off its power from the time the rainy season ends in autumn until the Okhotsk sea air mass starts to exert its influence in the spring of the following year. But in some years, the effect of the Siberian air mass continues almost to the onset of rainy season. That is, it lasts from late September till early May of the following year. Still, The most decisive effects of the Siberian air mass occur between December and February, with the most powerful effects in January. Throughout the entire time that the peninsula is under the spell of the Siberian air mass, the weather is cool, but the closer to midwinter, the colder.

Even in September, the air from the Siberian plains affects Korean weather. Although the air mass isn't powerfully developed at this time, when it does approach Korea, the weather suddenly cools down after being muggy the previous day.

With the coming of autumn, the weather suddenly cools down and the sky becomes clear blue. Once again, this is due to the influence of the air mass formed in the Siberian plain. The clear autumn weather is welcome by farmers as they harvest their crops. A rainy autumn is a curse: heavy rainfall can be disastrous for both rice and fruit harvests.

Even in early spring, Korea is under the influence of the air from Siberia.

The clear blue sky in early autumn The air from the Siberian plains brings clear weather to the peninsula, and helps the growth and harvest of crops. On such a day, one can get a clear view of Wolchulsan from Naju some 30 km away. (Naju, Jeollanam-do. Oct., 2007)

However, since the air is heated as the solar altitude rises and the continent becomes warmer, the air mass is not as strongly developed. This makes the cooling influence of the air mass less consistent. Thus, spring in Korea is typified by an unpredictable alternation of consecutive warm days and sudden cold spells. These intrusions of cold weather are called the 'spring frost' or 'final cold snap', and a poetic Korean expression suggests that they result from the cold's jealousy of the spring blossoms. The 'spring frost' appears at times when the Siberian air mass is relatively cohesive, but those phases do not last long. If you go outdoors during this time of the year, you will see flowers blooming here and there and new life sprouting in the valley streams, but snow still piled in the distant mountains.

The mountains and a stream around spring frost Flowers bloom here and there and new life sprouts in the valley streams, but snow is still piled in the distant mountains in early spring when the cold air from the Siberian plains brings a final cold snap. (Namwon, Jeollabuk-do. March, 2006)

Air from the Sea of Okhotsk

The Okhotsk sea air mass develops around the time when snow cover on mountains starts to melt and the cold water flows into the sea. So Korea is under its influence from late spring, when the snow cover starts to melt in the Siberia region. When the air mass from the Sea of Okhotsk approaches the peninsula, there is cool and clear weather throughout the country. Climb to any high place on such a day and you have wonderful visibility far into the distance. When you hear television news reports announcing the visibility of Mt. Songaksan, in Gaeseong far across the DMZ in North Korea, you can be sure it is a day when Korea is heavily influenced by the

The clear sky due to the effect of the air mass developed in the Sea of Okhotsk When the air mass developed in the Sea of Okhotsk approach the peninsula, the sky is clear even in the capital, Seoul. (Seoul. June, 2006)

Sea of Okhotsk air mass. That is to say, it is because a clean air mass is moving towards Korea from the East Sea. However, at such times it is cloudy and rainy on the east coast. At worst, it can be a very dark and gloomy there.

When the Okhotsk sea air mass is affecting Korea, the temperature is low to the east, but high to the west of the Taebaek Mountains. Such is the typical weather around the time of the *Dano* festival (May 5 by the lunar calendar) in Gangneung. Tourists arrive from the west side of the mountains dressed in midsummer attire only to encounter chilly weather in Gangneung that puts them at risk of catching a cold.

When the air that poured rain on the east coast passes over the Taebaek Mountains, its moisture is gone and conditions are dry. Prolonged dryness poses a danger to farming, so at such time sprinklers are used to water the fields.

Air from the North Pacific Ocean

The North Pacific air mass affects Korea's weather in the summer. This air is hot and humid. Therefore, during midsummer in Korea, your skin feels sticky even when you're sitting in the shade of a tree.

While it is moving towards Korea, the North Pacific air mass passes over the warm Kuroshio Current. During that journey, it picks up the water vapor evaporating from the ocean, and also becomes unstable. This instability is due to the convection that are activated by the great difference in temperature between the relatively cool air on upper layer and the air on surface, which has been heated by contact with the Kuroshio Current.

Under these unstable air conditions, ascending air currents are easily formed, and they develop cumulus here and there. If the cumulus are formed over mountain area, there is a shower in the late afternoon or early evening, which cools down the heat. These 'passing rain' showers are very brief, and give way to clear, sunny skies.

Thunderheads are formed from unstable air masses When the air is unstable, thunderheads are easily formed, and it showers or hails.

Further south, in tropical regions where summers are much hotter and muggier than in Korea, similar, but much stronger showers known as squalls fall at around three in the afternoon. Where drainage systems are inadequate, these squalls cause sudden overflows and flash floods. Adapting to such weather, people in those regions have long built houses raised above the ground by stilts. During squalls, it is pitch dark in the area immediately affected, but one can see the sun shining in the distance.

Under the influence of the air mass from the North Pacific, the Korean peninsula experiences a discomfort index as high as the soaring humidity level. Tempers fray, and conflicts are common. It may not be an exaggeration to blame the North Pacific air mass for many of the disputes that break out in such weather. But this midsummer heat also brings some benefits. Without that hot, muggy summer, rice cultivation would not be

Houses in the tropical regions In tropical regions, people have long built houses raised above the ground by stilts, which prevent the structures from being inundated by sudden flash floods, and the space between the ground and the floor of the houses cools the heat from the Earth. (Siem Reap, Cambodia. May, 2008)

possible. Since throughout its history, rice has been the staple food, Korea owes a great deal to the North Pacific air, although it affects the peninsula for just a short period of the year.

What causes seasonal rain?

The air masses from Siberia and from the Sea of Okhotsk are cold, while the air mass arriving from the North Pacific Ocean is hot. Therefore, where the cold northern air and the hot southern air meet, a long, wide strip of clouds develops, extending from east to west. Known as the polar front, what call the '*Jangma* front' this phenomenon accounts for a great deal of rain in Korea.

The polar front is positioned near or south of Korea throughout the year. When the mass of cold air from either the Siberian plains or the Sea of Okhotsk is strong, the polar front remains much further south than Korea. Conversely, when the warmer North Pacific air mass is the dominant force, the front moves further north. Sometime before midsummer, when the polar front reaches to Korea, the rainy season (*Jangma*) begins. The position of the front changes depending on the relative strengths of the cold air masses to the north, and the warm air masses to the south of the peninsula. Consequently, the rainy season is tediously prolonged as the polar front moves up and down the peninsula.

When two air masses of different temperatures meet, clouds form at their juncture by the same principle that makes water drops form on the side of a cold water glass on a hot, muggy day. The colder the water, the more water drops form. In the same way, the greater the difference in

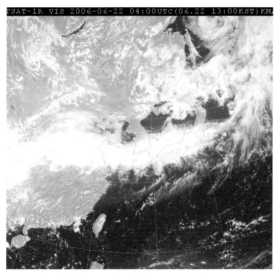

The formation of clouds when the polar front is in effect One can see a long horizontal band of clouds covering east to west of the peninsula. (Korea Meteorological Administration. June 22, 2006)

temperature between the air masses, the thicker the clouds that form where they meet. In other words, a bigger temperature difference between the two air masses leads to a heavier rainfall.

The rainy season lasts approximately one month, from late June to late July. In late August, when summer is yielding to autumn, the polar front is again positioned over the peninsula. Once again, the band of clouds moves up and down the country, dropping large amounts of rain. Flooding frequently occurs during this so-called 'late rainy season' because the water table is high and reservoirs are already filled with water from the rainy season earlier in the summer. Under those saturated conditions, it doesn't take much rain in the late rainy season to trigger serious floods.

As summer turns to autumn, the seasonal typhoons bring warm air up from the tropics. Accompanied by strong winds and heavy rains, the

When hot air meets with cold water When hot air and cold water meet, the steam pressure suddenly drops and congeals, making water drops on the side of a water glass on a hot, muggy day. In three glasses, ice cold water was poured half a glass (on the left), and a full glass (in the center), and normal water kept at room temperature (on the right). Water drops formed on the two glasses containing cold water, but no changes occurred for the glass on the right. By the same principle, when two air masses of different temperatures meet, clouds form at their juncture. (31°C, July, 8, 2008)

typhoons frequently cause extensive damage along the east and southern coasts of Korea.

Part 2

It's only really winter when it's cold; only really summer when it's hot

05

It has to be cold to be winter

We value a season most when the weather conforms to what the calendar dictates. We like our winters cold and our summers hot. In particular, the ideal of a consistently cold winter has real meaning for farmers growing barley. If a winter is warm, the result is not just weak, hypertrophied barley plants, but thriving infestations of diseases and pests that ruin the year's crop.

The he children living in the countryside used to look forward to the winter for the long vacation from school. Moreover, many exciting things happened around the time that the winter holidays began: the grand annual event of making winter kimchi; boiling soybeans to make soybean blocks (*meju*); and most of all, the preparation and consumption of red-bean porridge. This delicacy was traditionally eaten on the winter solstice, to ward off evil spirits on the longest night of the year. For children who normally subsisted through the year on barley, these were delightful occasions for tasting special delicacies.

The soybean blocks were hung to dry under the eaves all winter, then placed in a warm, heated room to ferment. Afterwards, in the following spring, the blocks were placed in earthenware pots to make soybean paste (*doenjang*), red-pepper paste (*gochujang*), or soy sauce.

Icicles and soybean blocks in the winter Once winter is over and spring arrives, it is time to make soybean paste using the blocks that were hung to dry under the eaves all winter. Icicles are hanging on the edge of the eaves. (Yangyang, Gangwon-do. Jan., 2003)

However, winter had its downside. In the staff-rooms of rural schools, pine-cones were burned for heating. Perhaps as a way to stimulate their students' competitiveness, the teachers had the pine-cones collected by each village. The older children pressed the younger ones to gather more pine-cones, truly a tedious chore. Back then, it was no small matter to get through the cold winter. The first time I visited Seoul, I was impressed by the sight of stovepipes sticking out of classroom windows. At that time there were no stoves in the classrooms on Jejudo Island. Nowadays, of course, with central heating, a stovepipe would seem laughably quaint, whether in the city or the countryside.

There are several things that come to mind when one hears the word 'winter': skating, sledding, young children running along rice paddies or hills, earthenware pots covered with snow, icicles hanging from eaves, memories of sipping hot noodle soup late at night after studying at the

Snow in winter Frozen rice paddies turn into a natural ice rink in the winter. (Cheorwon, Gangwon-do. Dec., 2007)

library, and eating roasted sweet potatoes with kimchi. All these are possible only when it's cold.

If it were to be warm in winter, there would be no sledding, no icicles hanging from eaves, and no layer of snow blanketing the world. It only feels like winter when the weather is truly cold. Today people worry that the winters are no longer so cold. They miss the necessary ingredient for the authentic experience of the season.

Why is it cold in winter?

In winter, there are fewer hours of daylight, so the earth's surface receives less solar energy. Naturally it gets cold. But that's not the whole story in Korea. During winter, this peninsula is covered by cold air blown in from Siberia, one of the coldest regions in the world.

During winter, the Siberian plains cool down very rapidly. On the other hand, the temperature drop is much more gradual in the waters of the Pacific Ocean. So, the air is much colder over the Siberian plains than over the North Pacific. Lower land temperatures cause an increase in air density as heavy air continues to accumulate near the ground. Higher air density is referred to as high air pressure, and the high pressure conditions of the air from the Siberian plains are a dominant feature of Korea's winters.

Air moves from areas of high pressure and low temperature towards areas where the atmospheric pressure is low and the temperature is warm. Thus, the cold Siberian air moves towards the warmer North Pacific region, and Korea is right in its path. And as the Siberian air mass moves across Korea, it brings along the strong, bone-chilling winds associated with the Siberian plain. Although the intensity varies according to region, at that time of year the whole Korean nation is in the grip of a fierce chill. When the temperature suddenly drops compared with the previous day, the Korea Meteorological Administration issues a cold wave warning. Recently, however, due to global warming, the cutting edge of the winter

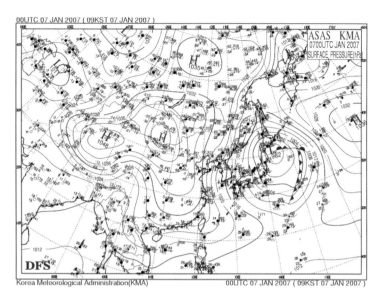

Korea Meteorological Administration(KMA)   00UTC 07 JAN 2007 (09KST 07 JAN 2007)

The surface pressure chart when the Siberian air mass affects Korea When the Siberian air mass moves towards the peninsula, there are high pressure anticyclones in the west, and low pressures cyclones in the east, with strong northwesterly winds. (Korea Meteorological Administration. Jan. 7, 2007)

chill seems to have been blunted. One rarely hears cold wave warnings anymore.

If cold air flowed in from Siberia through the entire season, the winter in Korea would be unbearably cold. Fortunately, long periods of continuous cold are rare. In order for an air mass to form, a certain amount of air must accumulate. It takes some time for this to occur on the Siberian plain. So, the chilling influence of the Siberian air mass usually affects Korea in a one-week cycle, rather than continuously.

When an air mass that has accumulated in the Siberian plains moves, it takes on the form of what meteorologists call an anticyclone. That is, a high pressure area forces air away from the center. That air is replaced in the center by a downward draft of air from higher altitudes, which becomes

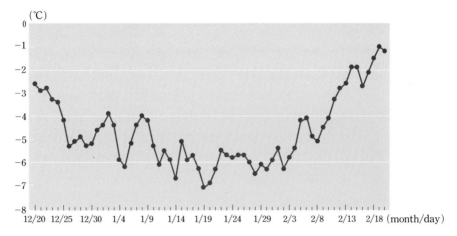

The pattern of winter weather changes An approximately weekly cycle produces the typical pattern of Korea's winter weather, traditionally described as "three cold days followed by four warm days." (Hongcheon, Gangwon-do. The average winter temperature of the years 1971 to 2000)

compressed and warmed as it moves down. This warming reduces the humidity of the descending air, which results in relatively cloudless, low-humidity conditions in the anticyclone. This anticyclone is a moving high pressure system that, once it reaches Korea, causes temperatures to rise considerably. Under its influence, Koreans enjoy days of comparative warmth in winter.

However, back in Siberia, it takes several days for another air mass to be transformed into an anticyclone. Consequently, there are gaps between the anticyclones that reach the peninsula, like valleys between mountains. These gaps are characterized by the low temperature and low atmospheric pressure of the untransformed Siberian air mass. The weather turns cold, and it rains or snows. Those conditions persist until the arrival of the next anticyclone from Siberia. This alternating sequence of influences from high pressure anticyclones and the low pressure gaps between them produces the typical pattern of Korea's winter weather, traditionally describe "three cold days followed by four warm days."

Today there are frequent claims in the media that the familiar three-cold-days-and-four-warms-days cycle has disappeared. According to this argument, it has been replaced by a cycle of two cold days followed by five warm ones, or a one-cold, six-warm-day pattern. But there is no substance to these claims. In earlier days, the inhabitants of the Korean peninsula, living lives that made them acutely sensitive to all aspects of the weather, recognized an approximately weekly cycle in the pattern of winter weather changes. They coined the phrase, "Three cold, four warm" to capture that general observation. The key is the general weekly periodicity, not the exact distribution of days. Strictly speaking, the three-four pattern has never been exact, neither for our ancestors nor today.

Hangang observation point for ice formation The yardstick for judging whether Hangang is frozen is seeing whether ice is formed on the river 100 or 200 meters upstream between the third and the forth beams (southbound) of Hangang Bridge, which courses are the waterway for sightseeing boat cruises. (Hangang Bridge, Seoul. Jan., 2008)

There has been much discussion of global warming of late. There are many reasons for this heightened awareness and concern, but it is true that Korean winters now are considerably warmer than in the remembered past, especially the 1960s. Koreans who were students back then remember many winter days when schools were closed due to cold weather. Nowadays Korea rarely has a snowstorm severe enough to close schools. Although we still use the phrase 'freezing cold winter', it hardly suits the current reality. Only rarely does the Hangang freeze. In the 1970s the Hangang would start to freeze from late December. But today, if the Hangang freezes at all, it is only in the dead of winter, after mid-January.

The authentic feeling of winter

Do we really need extremely cold winters? Today, probably not too many people think so. However, there is something valuable about having different seasons marked by intensely different weather. Hot or cold, the inhabitants of a region have adapted to handle each season.

In Korea, the idea that it is best for winters to be truly cold seems to be connected to the cultivation of barley as a winter crop. A very cold winter, with plenty of snow is optimal for barley farming. If the winter is warm, barley plants grow too large. In the event of a spring frost, such hypertrophied plants are unable to adapt to the cold, and will freeze to death. A covering of snow in the field insulates the barley from the cold, and when it melts in the spring, it provides a suitable amount of moisture. This moisture is especially important in the spring, when the barley reaches its full growth. Apart from the melted snow, cover that is the driest time of year. As a child, I often heard farmers say that a good crop of barley could only be grown if snow cover remained on Baengnok-dam (White Deer Lake), at the summit of Hallasan. Moisture from that melting snow cover nourished the barley fields in the lowlands. Warm winters pose another danger to barley cultivation: diseases and pests that aren't killed off by

Barley fields in winter Snow is optimal for barley farming. A covering of snow in the field insulates the barley from the cold, and when it melts in the spring, it provides a suitable amount of water. (Gimje, Jeollabuk-do. Dec., 2005)

freezing can ruin a crop.

In winter, Korean farmers wrap bands of straw around tree trunks. The aim is to protect trees from the cold, and also to control insects and pests. Instead of attacking the tree, these gather inside the straw during the cold winter. When spring comes, the farmers remove the straw and burn it, along with the insects and pests. But this wonderfully organic way of controlling pests is only effective if the winter is cold. In a warm winter, the pests have no incentive to seek the warmth of the straw.

Farmers are not the only people who benefit from cold winters. Merchants of winter garments such as fur coats depend on low temperatures to sell their products; the same is true for companies that sell heating equipment. Ski resorts do good business when there is plenty of snow piled up. Unless the weather is cold, the snow will melt and they will need to spray the

Bands of straw are wrapped around young trees to protect them from the cold in winter, and to control pests in early spring. (Yeongju, Gyeongsangbuk-do. Feb., 2008)

slopes with artificial snow: an expensive process that reduces profits. But outside of such businesses, ordinary people have no great fondness for cold winters. And as fewer and fewer farmers grow barley, the ideal of truly cold winters may soon be a thing of the past.

06

The joys of spring

In spring, the heart leaps as flowers bloom in the mountains and fields. Men and women, young and old, everyone feels the excitement. Korea's spring weather is also very diverse, with spring frosts, droughts, northeasterly winds, and yellow dust. This diversity is due to the variety of meteorological influences that affect Korea at this time of year.

n spring, the heart leaps as flowers bloom in the mountains and the fields. Men and women, young and old, everyone feels the excitement. The first image that comes to mind with the word 'spring' is a beating heart: the rapid, eager pulse of youth. It may be that people feel the excitement of liberation after being confined by the cold all winter.

The thought of spring also brings to mind the image of a countryside filled with flower blossoms. Indeed, venturing out to the fields in the Korean spring, you find a world of colorful flowers. Some countries have acquired fame for their spring flowers. Prime examples are Daap-myeon in Gwangyang, Jeollanam-do, and Sandong-myeon in Gurye, Jeollabuk-do along the Seomjingang. Daap-myeon is known for white Japanese apricot blossoms, and Sandong-myeon for yellow Japanese cornel dogwood, whose flowers cover the whole village in spring. Hwangjangjae Pass, between

White Japanese apricot blossoms announce the arrival of spring in the southern provinces in Korea. (Gwangyang, Jeollanam-do. March, 2006)

A farmer and his ox work hard, plowing a new field in spring. (Yeongwol, Gangwon-do. April, 2008)

Andong and Yeongdeok, is full of peach blossoms when the flowers are at their peak. Take one step away from the throng of tourists in these areas and you'll find a hard-working farmer out in the fields with his ox, and hear the ringing of the ox's bell. Spring has truly arrived. Some green grass is starting to appear in the dry, yellow fields.

There are many things that carry the scent of spring even in areas near the cities. Housewives gather by riverbanks or low hills where there is some green grass growing, and together they pick spring herbs. This is also a sign of the arrival of spring and of people breaking free from the cold confines of winter.

Korean spring weather: like a woman's heart?

Korean folk sayings liken spring weather to a woman's heart. The analogy is based on the idea of capriciousness. But is Korea's spring weather really capricious, or just diverse? Back when I was a military cadet in Daejeon, the training officers used to say to the trainees, "Daejeon's weather is like a woman's heart." I understood them to mean that it was subject to unpredictable changes, so many times I looked up at the clear morning sky and foolishly expected it to rain. But Korean spring weather is not that capricious.

Back when I was serving as a weather officer in the Air Force, spring wasn't a season I looked forward to. We were required to send out the first weather forecast of the day before 5 a.m., based on a weather map at 3 a.m. In those days, our weather maps were pretty rudimentary. A weather soldier received the data by telegraph, and then manually recorded the information on a map. Mistakes in the telegraphed data were corrected by phone calls from one base to another, as we fixed numbers and filled in

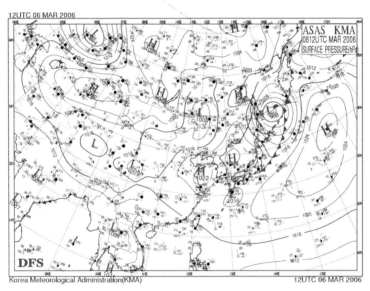

The surface pressure chart of Korea and her surrounding region in spring The disposition of atmospheric pressures around the peninsula in spring is extremely complex. (Korea Meteorological Administration. March 3, 2006)

blanks. So it wasn't until 4:30 a.m. that a completed weather map was finally put in the forecaster's hands.

The problem started there. The complicated map was not easy to analyze, and once it was past 5 a.m., the phone wouldn't stop ringing with urgent calls for the weather forecast. Under that pressure, the forecaster labors nobly to analyze the map, but making sense of the complicated isobaric lines is no easy task. The disposition of air pressures around the peninsula in spring is extremely complex. In winter, on a day when the pressure systems follow a simple pattern, the weather map would be completed based on just five or six data-points for High and Low. But in spring, there were dozens of pressure systems printed on the map. And still, even after all that data was included, the map was bound to contain inaccuracies.

With such difficulties in just making a weather map, it was doubly hard to come up with a forecast. Spring is the most difficult season for judging which of the pressure systems positioned near Korea is most likely to affect the weather. Consequently, our forecasts often proved wrong.

This variety of pressure systems and air coming from different directions accounts for Korea's diverse spring weather: spring frosts, droughts, northeasterly winds, yellow dust, and sea fog and so forth. First of all, the Siberian air mass that so heavily influences Korea's winters continues to affect the peninsula even in spring. However, its force is considerably weakened, so its characteristics are less stable. Although it's the same Siberian air mass, in its modified state, it influences the weather differently. Furthermore, in late spring, the Sea of Okhotsk air mass starts to expand its force, frequently passing across Korea with its associated warm temperature and low atmospheric pressure.

Migratory High from Siberia usually affects Korean weather in spring, but in early spring, the Siberian High also temporarily gains strength. That is when it feels like winter again. Especially from late February, after several

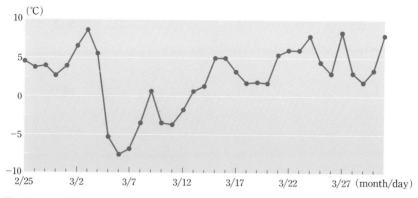

The changes in temperature in spring Temperate early-spring weather suddenly drops below freezing point when the infamous spring frost hits Korea. (Seoul. March, 2007)

warm days, the spring frost hits the country in early March, at the start of the school year. Parents have to dress their children as if for the middle of winter, and pediatric clinics are suddenly crowded with young out-patients.

Two Korean proverbs refer to these spring conditions: "The kimchi pot cracks in the February wind," and, "The prematurely old man freezes to death in the spring frost." February here refers to the traditional lunar calendar, so is roughly equivalent to March by the solar calendar we now use. The proverbs testify to the ferocity of the spring frost. It is said that the spring frost is due to the jealous Wind God, who is trying to stop flowers from blooming. This cold spell usually comes between late February and early March (on the solar calendar). It is frequently reported in the media when the spring flowers, namely forsythias and azaleas, are about to bloom, from mid-March to early April (also on the solar calendar).

The spring frost not only delays the blooming of flowers, but also affects the farming schedule. If the abnormally low temperatures associated with spring frost appear in late spring, the result can be cold-weather damage to crops that have started to sprout early. Such late frosts are deadly to young seedlings. Agricultural-related institutes monitor the weather for the

The first day of school Parents have to dress their children as if for the middle of winter, in early March, at the start of the school year. (Seoul. March, 2003)

approach of a cold spell, and issue warnings to farmers to keep their crops warm. But for crops planted in outdoor fields, not greenhouses, there is nothing to be done. So, farmers must refer to climate records from years past in order to determine the best time to plant their crops.

The fires of spring

Somehow, forest fire seems to have become the Herald of Spring, making its appearance around this time each year. Far beyond the scale of a small mountain fire that concerns a single village, this is a fire that worries the whole nation. The frequent forest fires along the east coast of Gangwon-do most often catch the media's attention, but during the spring forest fires can break out in virtually any region of the country. Sometimes there are so many fires burning simultaneously throughout the nation that there aren't enough fire helicopters to head out to all the sites. The trees and bushes are completely dried out after the long, dry winter, and as the temperature rises, the relative humidity drops. Under those conditions, without proper precautions, even a small blaze can turn into a massively devastating forest fire.

The migratory High affecting Korea in the spring move rapidly from west to east. That is why the weather in spring changes so quickly. After a High has passed, there is a extratropical cyclone, followed by a visit from another migratory High. Such is the typical pattern of Korean spring weather. Around this time the weather isn't so much changing as dawdling. What looks like impending rain fails to materialize; the result is spring drought. This is the time when accounts of forest fires appear in the media.

During spring afternoons, when the air is dry, there are great differences in the surface heating of different regions. The resulting temperature differences stir up strong air currents. So, sudden gust of wind are common

Forests in spring The trees and bushes are completely dried out after the long, dry winter, therefore, without proper precautions, even a small blaze can turn into a massively devastating forest fire. (Jangsu, Jeollabuk-do. April, 2008)

on spring afternoons. If you happen to be driving across a bridge during one of these gales, you can feel your car being buffeted from side to side.

And these spring gales are a major culprit in the spread of forest fires. The reason forest fires grow so massive in the region east of the Taebaek Mountains is the spreading effect from those strong winds. Moreover, since the wind blows from the west, the air is much drier on the other side of the mountains. This relative dryness also contributes to the rapid and extensive spread of forest fires in the region east of the Taebaek range.

Mountains stripped bare by forest fires have become a familiar sight on the east coast of Korea. A forest fire instantly destroys trees that have taken several decades to grow, and even more seriously, by removing the tree cover, it exposes the earth to rain. So, on a mountain that has suffered a forest fire in the spring, heavy rainfall can lead to mudslides in summer.

Mountains stripped bare by forest fires have become a familiar sight on the east coast of Korea. (Samcheok, Gangwon-do. Aug., 2001)

The yellow dust season

Around the same time that there are frequent forest fires, Korea often plays host to another unwelcome spring guest: yellow dust. This consists primarily of minute particles of fine dry soil blown from the Huangtu Plateau in China or from the Gobi desert. When a strong wind or low passes that area, a dust of these yellowish particles is blown into the air by ascending air currents. Strong winds in the upper atmosphere transport the yellow dust all the way to Korea. The stronger those winds, the further they will carry the yellow dust, which appears not only in Korea, but also Japan, and even as far away as Hawaii.

Yellow dust sharply reduces visibility, and is a particular nuisance similar to the nuisance of smog. The minute dust particles can damage delicate equipment, and even worse, some of the dust's components are harmful to the human body. Therefore, when there is for yellow dust in Korea, the

Another unwelcome spring guest: yellow dust When a strong wind blows in the Huangtu Plateau in China or in the Gobi desert, the upper atmosphere transport minute particles of fine dry soil, the yellow dust, all the way to Korea. (Seoul. April, 2006)

A forestation project in the Huangtu Plateau area in China Korean non-governmental organizations are conducting campaigns to plant trees in the Huangtu Plateau, as an effort to prevent yellow dust. What looks like thin sticks in the image are the newly planted trees. If one walks on the ground, it feels like treading on fine sand. (Linzhou, China. June, 2007)

meteorological agency issues warnings to avoid outdoor activities. In especially severe situations, schools are temporarily closed.

The yellow dust invasion of Korea has intensified in recent years. Some studies have argued that this is also a consequence of global warming, and predicted further intensification as temperatures rise worldwide. Korea has joined with China and Japan in efforts to combat the problem, and Korean non-governmental organizations are conducting campaigns to plant trees in regions where the yellow dust originates. In the Huangtu Plateau region in China, steady efforts are being made to prevent yellow dust.

Even in late spring, it is cool on the east coast

Around the time of year when the yellow dust ceases to appear in Korean

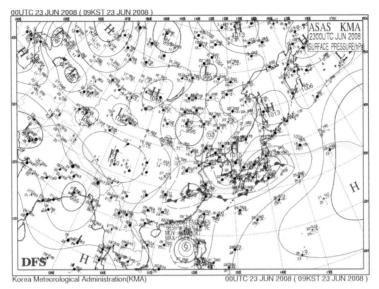

Korea Meteorological Administration(KMA) 00UTC 23 JUN 2008 (09KST 23 JUN 2008)

The surface pressure chart when the Okhotsk air mass affects Korea When the air blows from the Sea of Okhotsk and influences Korea, a strong high-pressure anticyclone develops in the northeast of the peninsula and prevailing wind starts to come increasingly from the northeast. (Korea Meteorological Administration. June 23, 2008)

skies, the northeasterly starts to come increasingly from the Sea of Okhotsk. Although it is now late spring, the Sea of Okhotsk is still cold, so the northeasterly wind feels quite cool. During this period, the weather remains cloudy and overcast in the region east of the Taebaek Mountains. For example, in Gangneung, daily maximum temperatures rarely exceed 20°C, so visitors from other parts of the country feel chilly. By this time of year, there are no more reports of forest fires.

When the northeasterly wind blows, the visibility is superb in the region west of the Taebaek Mountains, and the clear skies are refreshing to the inhabitants. Condensed water in the air is vaporized as it passes over the range, so the skies are clear. But looking towards the east, one can see layers of clouds reaching higher than the Taebaek Mountains. This is the time of

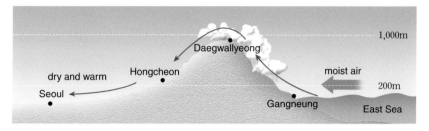

The humid air from the East Sea releases its moisture as rain as it crosses the Taebaek Mauntains, so when that air descends on the leeward side, in Hongcheon, the air is warmer and dryer than on the windward side, in Gangneung. This warm and dry northwesterly is called *nopse* or foehn.

year when you can see beautiful, clear white clouds floating in the bright blue sky.

There is a big difference in temperature between the regions west of the Taebaek range and the regions to the east. This is due to the difference in temperature between the moisture-laden air that rises to the top of the mountain on the windward side, and the air that descends on the leeward side after losing its moisture.

When the northeasterly wind is blowing, the temperature normally rises above 30°C in the west regions of the Taebaek range, and it is also dry because the clouds have released their moisture as rain while they were crossing the mountains. This is a so-called *Nopsae* which is the warm and dry northeasterly that blows in the regions west of the Taebaek range. But even with the temperature over 30°C, it isn't uncomfortable, thanks to the low humidity. You feel hot under the sun, but cool in the shade. In the absence of humidity, you don't feel sticky, and the heat quickly dissipates from your skin.

Pleasant as it is, if the northeasterly wind continued too long, prolonged dry conditions could lead to a crop-damaging by drought in the west region of the Taebaek range, it could suffer cold damage in the east coast of Gangwon-do. That is why in the east regions of the Taebaek range, most

Rice farming in the east of the Taebaek Mountains Because the regions east of the Taebaek Mountains could suffer frost damage from the Sea of Okhotsk air mass, most farmers plant quick-ripening strains of rice, as well as rice that take longer (left in the photo) to ripen. (Gangneung, Gangwon-do. Aug., 2008)

farmers plant quick-ripening strains of rice that can be harvested early in the season. This is the sort of rice usually grown in the highlands. In the lowlands of other regions of Korea, farmers plant rice that takes longer to ripen, and harvest it later in the season.

07

Why do we need the rainy season?

With all its discomforts and inconveniences, one might wonder whether there are any Koreans who actually welcome the rainy season. But without it, the rice crop would be at serious risk. We depend on that rain to wet all the mountains, streams, trees and grass, and to fill the reservoirs to provide an ample supply of water for farming.

As one season ends, most people feel wistful, but they also look forward eagerly to the coming of the new season. The anticipation of novelty is exciting. But does anyone feel such eager anticipation at the approach of the rainy season (so-called *Jangma*)? One might doubt it, so unwelcome is the rainy season in Korea.

Nor is there anything very pleasantly promising about the images evoked by the phrase 'rainy season': floods, damp, humidity, tedious days, and so forth. Whether it rains a lot or a little, if the rainy season goes on for too long, the farmer's worries increase. Along with the rains come more gray days with high levels of humidity. This makes the crops increasingly vulnerable to disease. The outbreak of a plant disease can spell disaster: a whole crop can be ruined as the disease spreads from farm to farm across a large region.

Chili blight spread during the rainy season With continuous days of high levels of humidity, crops, such as chili, become vulnerable to disease. (Naju, Jeollanam-do. Sept., 2007)

On the other hand, floods are quite rare during the rainy season. Rain and floods are two separate matters. In the rainy season, it just rains a lot. Of course, a torrential rain would cause a temporary inundation, but real flooding only occurs when soil is in saturation and the reservoirs are already full. Neither is the case at the start of the rainy season in Korea.

What would Korea be like without its rainy season? First of all, rice farming would be imperiled. Right before the rainy season, Korea experiences a period known as the 'drought before the rainy season'. In late spring most of the reservoirs that are dug throughout the agricultural countryside are dried up. Looking into such reservoirs, the bottom is revealed, the earth dried and cracked like a turtle's shell.

Small streams also dry up to reveal bare ground. It is up to the rainy season to refill those streams and reservoirs. And it is thanks to this rain

A small riverbed right before the rainy season If the dry period before the rains is prolonged, small reservoirs or rivers reveal the bottom. (Yeongdong, Chungcheongbuk-do. May, 2008)

that farmers can grow rice in Korea. At first thought, it may seem that nobody would actually be eagerly awaiting the rainy season, but the urgency of the situation is obvious in the worried expressions on farmers' faces when the 'drought before the rains' is prolonged. In 1994 there was hardly any precipitation in the summer. The fields across the country were literally drying up. Ordinarily forests fires are exceedingly rare in the summer, but the dry summer of 1994 saw forest fires on Jejudo Island.

Why does the *Jangma* front move up and down the peninsula?

Since Korea is located in westerly zone, meteorological phenomena generally move across it from west to east. But unlike other weather patterns, the polar front follows a north-south path. What's more, it doesn't move steadily in one direction, but shifts north and south, back and forth, ranging up and down the peninsula. Maybe for that reason, the forecasts for the onset of rainy season are often incorrect, although the general accuracy of weather forecasting has greatly improved in recent years.

Jangma are carried by the polar front that is formed where the two air masses from the cold regions north of Korea meet the warm one from the southern region. Although it is summer, it is cool north of that front, and warm to the south. Therefore, although it's called 'rainy season', the actual weather depends on the current location of that front. As long as it is south of you, your weather is cool; as soon as it passes to your north, you experience warm, muggy conditions. Along the front itself, it often rains heavily. So, like the famously capricious weather of a Korean spring, rainy season weather is also quite variable from one day to the next.

Where the rainy season differs from other seasons in Korea is in the regional variability. During the other seasons, whatever the weather on a given day – rain or shine, wind or snow – conditions are fairly uniform across the whole country. However, in the rainy season, on the same day, it

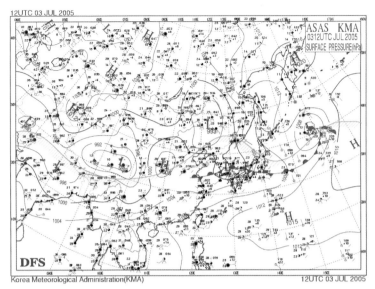

ASAS KMA
0312UTC JUL 2005
SURFACE PRESSURE(hPa)

DFS

Korea Meteorological Administration(KMA)   12UTC 03 JUL 2005

When the ploar front has reached Korea The ploar front follows a north-south path and affects the weather in Korea. (Korea Meteorological Administration. July 3, 2005)

can be muggy in some regions, and cool elsewhere. Or there can be a torrential rain in one quite restricted area, with sunshine on the rest of the peninsula. This regional variability makes the rainy season the most difficult time of year for weather forecasters.

The weather of the rainy season is not unique to Korea. One can experience a similar type of weather anywhere in East Asia that is simultaneously affected by the air masses of Siberia, the Sea of Okhotsk, and the North Pacific. Called '*maiyu*' in China and '*baiu*' in Japan, the rainy season (*jangma*) weather is the same, although it falls at a slightly different time. And as in Korea, the front ranges across those regions, north and south, back and forth.

On a first day of the rainy season in Korea, the front usually starts from the south and moves up north. In a typical year it hits Seowipo, on the southern side of Jejudo Island, around June 20. The rain usually reaches Seoul around June 25. The end of the rainy season, however, comes to both places at nearly the same time. Therefore, even if early-bird college students in Seoul rush back home to Jeju as soon as their summer vacation starts, hoping to save on travel expenses by making the trip to the island before the peak tourist season, they end up following the seasonal rain, garnishing the finale by bringing the rain front with them to Seoul a week or so later.

What time the rainy season begins is a matter of crucial importance both to farmers and to the people in charge of managing water resources. If the rainy season is delayed, the water level in the reservoirs drops, and a state of emergency is declared. After the spring planting, the rice paddies need plenty of water. But if the rainy season is late to arrive, the reservoirs are empty. Thus, a prolonged delay of the rainy season can spell disaster for the rice crop.

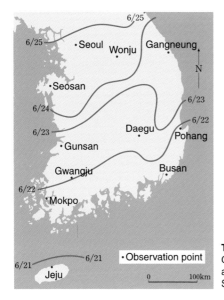

The ploar front chart
Generally, the front starts from the south around June 21 and moves up north, reaching Seoul around June 25.

Water is a serious matter in Korean farming, as evidenced by the Korean proverb: "Even between a father and a son, nobody yields a drop when it comes to apportioning water during the rice-planting season." A delayed rainy season is especially alarming if it comes after a long spring drought. One year when the rainy season was late, one farmer tried to take water from a neighbor's rice paddy to irrigate the rice he was planting. He was arrested and charged with larceny. That's how precious water is around this time of year.

Rain is indeed welcome in the early stages of the rainy season. But too much precipitation towards the end makes it difficult to manage water. When a torrential rain falls near the end of the season, there is high probability of flooding. The government manages water by building huge, multi-purpose dams, such as the Soyanggang and the Chungju dams, on

Water discharge from Soyanggang Dam Should the water released from a dam reach the lower reaches of a river after torrential rains, flooding is likely. It is quite rare that huge, multipurpose dams discharge water. (Chuncheon, Gangwon-do. July, 2006)

the main rivers and streams, but still, water management is not an easy matter. When it is certain that it will rain a lot, the dams must be emptied. But once a dam is empty, if the expected rainfall fails to materialzie, water shortage can result. But the government can't just allow the dams to remain full. With a full dam, it would be necessary to open the gates quickly to release excess water in the event of heavy rainfall. But any miscalculation in the timing or amount of such a discharge can have catastrophic consequences. Should the water released from a dam reach the lower reaches of a river during high tide, flooding is likely. Such are the complexities of water management that Korea continues to suffer many floods, despite the multi-purpose dams that are still being built.

A premature end of the rainy season can mean a summer-long water shortage. But once the reservoirs and dams are filled, that ceases to be a concern. From that point on, the worry is that the rainy season will continue too long. The extra rainfall from a prolonged rainy season can lead to flood damage. Moreover, diminished hours of sunlight can hinder the growth of crops. In the past, even a slight shortening or lengthening of the rainy season caused farmers to sigh and everyone else to worry about the price of rice.

Turning on the heater in summer

The rainy season brings more days of rain than other seasons. The sun does come out occasionally, but most of the time it is either raining or cloudy. So, the humidity is high throuoghout this season, which means that it feels very damp indoors, even when it isn't raining. Moreover, when the polar front is positioned south of the peninsula, the northern air mass exerts its influence, so conditions are not just humid, but cool and dark into the bargain.

When I was in middle school, we students were required to change to our summer uniforms from the first day of June. But that is right before the rainy season, when the weather is often chilly. Many days were too cold for short pants and short-sleeve shirts, so goose pimples were common. Using the weather as an excuse on those days, some of the senior students would walk to school in long pants. Perhaps their goal was to look attractive and mature, but in any case, they wanted to wear long pants. Teachers in charge of student conduct would be stationed by the front gate, on the lookout for students in long pants. Scissors in hand, the teachers were ready to cut the trousers of any offenders. It was the same hide-and-seek every year at the start of the rainy season.

Actually, on such a day, at home, one suddenly misses the heater. On a rainy day in the summer, residents of centrally heated apartment buildings wish that the management would turn on the heat. But there's not much chance that an office run by fixed rules will grant that wish. If only the apartment managers would save all the excess heat they provided in the winter, when residents have to open their windows because it's so stifling indoors.

It would be perfectly reasonable to turn on the heat at this time of year. The humidity is so high both indoors and out that the laundry hardly dries, and the temperature is a good 10 degrees lower than in midsummer. It feels like the weather east of the Taebaek range, just before the rainy season. Turning on the heater not only warms up the house, it also cuts down the humidity. If the heat isn't turned on, molds grow in profusely, so if the humid weather persists, the news starts reporting incidents of food poisoning. So, running the heater is important to maintaining good health in the wet summer season. The rainy season lasts longer on Jejudo Island than elsewhere in Korea, which may explain why houses on Jeju were traditionally built with no chimneys in the kitchen. The heat and smoke from the kitchen stove would have been some help against the prevailing damp conditions indoors.

The year I moved to Seoul, after living for a few years in the

The kitchen of a traditional farmer's house on Jejudo Island Houses on Jeju were traditionally built with no chimneys in the kitchen. The heat and smoke from the kitchen stove would have been some help against the prevailing damp conditions indoors. (Seongeup Folk Village, Jeju-do)

countryside, I came to live in a basement apartment due to the high price of real estate in the metropolis. People who have never lived below ground level may not understand this, but it's a place where I never want to live again. The worst was during the rainy season. I never knew when water might come gushing in; with mold growing everywhere I was constantly worried about what went into my baby's mouth; and the washing would never dry. None of these are fond memories to recall. The landlords of such damp basement dwellings should at least waive the cost of fuel for the heater during the rainy season.

08

The sweltering summer is here

Korean summers are swelteringly hot. Just as winters must be cold, for the authentic experience of the season, summers must be sizzling hot. If the summer isn't hot, it's a lean year for farmers. In addition to the heat, the showers that temporarily cool the sweltering heat of midsummer afternoons are a distinctive feature of Korea's summer weather. Just as farmers in a drought look forward to rain, people look forward eagerly to those summer afternoon showers.

ummer is meaningful to Koreans in many respects. Thanks to the summer season, rice cultivation is possible on the peninsula, earning Korea's inclusion in the rice-farming cultural sphere. Throughout Korean history, many elements of life have been intimately tied to rice-farming. So without its hot summers, Korea would have a completely different culture.

The word 'summer' traditionally brought to mind the image of green

A green rice field in the summer. (Gimje, Jeollabuk-do. July, 2006)

The sweltering summer heat draws many people to the beach. (Gangneung, Gangwon-do. Aug., 2007)

rice fields, but nowadays most young people probably imagine a crowded beach instead. Many people, young or old, look forward to the summer season. Most likely that is because of the prospect of a vacation trip to the mountains, valleys, or seaside. On the other hand, for some people, the mention of summer evokes images of rain showers.

It may have to do with the well-known Korean novel, *Sonagi* (literally "The Shower (of Rain)") by Hwang Sun-won, but even without the literary association, many Koreans would link summer with rain showers. They are truly emblematic of summer weather in Korea.

Other images come to mind: crowds of people gathered along riverbanks at night to avoid the heat; the lookout shed in the center of a field where melons are ripening, old men playing a game of chess in the shade of a zelkova (a tree of the elm family that grows in East Asia) or the great floods following a sudden torrential downpour. Showers and lookout sheds are synonymous with summer. Growing up in a small rural village, I had to stand guard in a shed during the summer, keeping watch over the fields. So there's not much nostalgia in my memories of those sheds but the recollection of childhood boredom. A summer day felt terribly long for a child. Yet, even as a young boy, I felt a certain wistful emptiness as I walked down the hill and back to my home on the last day of lookout duty in the shed. Even as I anticipated the abundance of autumn, I somehow couldn't escape the feeling that something was missing when the summer ended. And it wasn't just because the summer holidays were over. As a child, the bigger reason would have been the thought of no longer being able to eat watermelons, the only dessert at that time.

Meanwhile, I still have fresh memories of the rain showers, and my fear

A lookout shed at a park Today the only lookout sheds remaining are those by state roads where farmers sell fruits, or shelters made to feel some nostalgia. (Guri, Gyeonggi-do. Oct., 2007)

as they approached the lookout shed from Hallasan. Spending all day looking at the dark clouds, I felt like I had a sixth sense for when the clouds would reach our village, accurate to the minute. The wet from these showers would penetrate right through the flimsy shed. But thanks to that experience, I acquired an acute eye for showers approaching from afar. I'm always the first to notice them.

Why is it so oppressively hot in the summer?

Korean summers are swelteringly hot. Just as winters must be cold, for the authentic experience of the season, summers must be sizzling hot. If the summer isn't hot, it's a lean year for farmers. So the intense heat of summer is vital: not for the sake of young people enjoying the beach, but for the nation's economy.

The summer's sweltering heat helps Korea's staple grain, rice, to grow, and makes the grass and trees flourish. Left alone during the summer vacation, the playgrounds of rural schools were overgrown with weeds. On the orientation day before the start of the next term, pulling the weeds was the major task. The front yards of our houses also required weeding. This important chore was the responsibility of the children who were too young to help out in the fields. At the slightest negligence, the family garden would be overgrown with weeds in the summer.

On Jejudo Island, where the weather is more hot and humid than on the Korean mainland, clearing the weeds from the family burial grounds is another important chore. Conscientious families tend the grass once or twice before August, but even so, by late August the grounds are overgrown with weeds and look like abandoned sites. Seeing the apparently neglected tombs, visitors from Seoul might rebuke the islanders for failing to show respect for their ancestors by tending their graves. That is why Jejudo Island has something called a weeding holiday. On Saturday mornings in early September, Gimpo Domestic Airport in Seoul is crowded with Jeju

Even well-tended family graves look like they have been neglected on Jejudo Island because it is more hot and humid than on the Korean mainland. (Jeju, Jeju-do. Aug., 2006)

natives flying down to their hometowns to cut the grass and to tend their family graves.

We loosely refer the months of June, July, and August as the summer season. However, strictly speaking, true summer only begins with the end of the rainy season. The weather during the rainy season is completely different from the weather that follows. The sweltering heat starts in earnest after the rainy season, and shows the true face of summer. From this time, the hot, muggy air, heavily laden with water vapor, starts to flow in from the North Pacific, where such sultry conditions prevail all year round. In winter, this North Pacific air cannot approach the Korean peninsula, which is covered by air from the Siberian plain. But as the solar altitude rises and the continent starts to warm up, the Siberian air mass loses its force. Then the polar front that develops between the two air

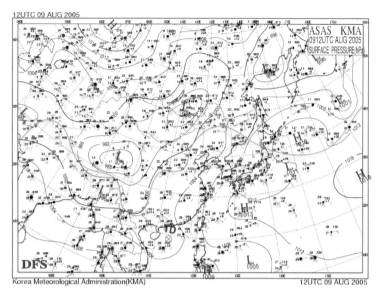

The surface pressure chart in summer The North Pacific Ocean high pressure anticyclones affect Korea in midsummer, developing high pressure air masses in the south and low-pressure air masses in the north. (Korea Meteorological Administration. Aug., 9, 2005)

masses slowly starts to move northward. But finally, after dumping seasonal rain on the peninsula for nearly a month, the front between the two air-masses moves beyond Korea and it is the air from the North Pacific, unopposed, that dominates Korea's weather. Now it is midsummer.

A descending air current develops where the North Pacific air mass is formed, so the weather is clear and sunny. Consequently, the region receives more solar energy than regions at other latitudes, even those closer to the equator. Furthermore, because of the vast ocean below, the air is laden with water vapor. Hence the North Pacific air mass has both high temperature and high humidity throughout the year. Naturally any area that comes under the influence of this air will experience hot, humid weather.

Under the influence of that hot, humid air mass, Korean summers are characterized by a high discomfort index. When this index exceeds 80, more than 50 percent of Koreans say they feel uncomfortable due to the weather. Since it has such an effect on daily life, the Korean Meteorological Administration's forecast of the discomfort index is followed closely throughout the country. When it is high, people turn on their air conditioners or electric fans, with a sharp increase in power consumption that makes the Korea Electric Power Corporation worry about its reserve capacity.

The three factors that affect the discomfort level are the temperature, humidity, and wind. Even when the temperature is high, if there isn't much water vapor in the air, the weather is bearable. For instance, before the rainy season, conducting a class in a lecture hall isn't a problem even when the temperature surpasses 30°C. But in July, after the rainy season is over, if the temperature exceeds 30 degrees, it's extremely uncomfortable to manage a class. The difference is the humidity. When the humidity is high, it is hard for moisture to evaporate from your skin, so you feel uncomfortable. In that situation, some wind, by aiding that evaporation, will reduce the discomfort level.

Going back just a few decades, the images of summer are completely different from the crowded beaches of today. Would I be showing my age if I admitted to remembering the sight of elderly people dressed in ramie hemp clothes gathered under the pavilion tree by the village entrance, leisurely fanning themselves? That may have been the best way to overcome the sweltering Korean summer.

As a gift for friends from abroad, many Koreans buy fans with the Taegeuk pattern from the Korean national flag printed on them. These gifts are much admired and enjoyed by their non-Korean recipients, but most of them are surprised when they learn the actual purpose of the fans. The fans testify to the extreme mugginess of Korean summers. A fan doesn't create a temperature change, yet you have the feeling of a cool

A pavilion tree by a village entrance Old villages still have a pavilion and a big tree where people gather under its shade in the hot summer. Functioning as the village square, on Jejudo Island, this pavilion tree is generally in the center of a village. (Yecheon, Gyeongsangbuk-do. Aug., 2007)

breeze when you fan yourself. This is because the flow of air from the fan helps to evaporate the humidity from your skin more quickly. It is similar to the method of cooling a patio on a midsummer afternoon by spraying it with water and letting it evaporate: the evaporating water takes away the surrounding heat with it. An electric fan, of course, follows the same principle as the hand-held paper fan: simply by putting air in motion, it achieves a cooling effect.

When the humidity is high, the temperature does not drop even at night. That's because the water vapor retains the heat. Nights when the temperature never falls below 25°C are called tropical nights. Such nights are so uncomfortable that it's difficult to fall asleep. Therefore, many people gather by riverbanks to avoid the heat. The discomfort frays

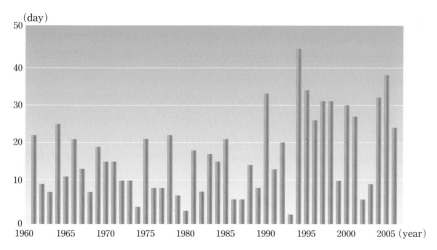

The change in the number of tropical nights on Jejudo Island Global warming and urbanization are combining to increase the number of tropical nights in most parts of the peninsula. (From 1960 to 2005)

tempers, and people quarrel over trivial matters. Conflicts break out that would never occur on a cold winter day. Global warming and urbanization are combining to increase the number of tropical nights in most parts of the peninsula. If the trend continues, we may reach the point of suffering through tropical nights throughout the entire summer.

Why are there so many rain showers in summer?

Rain showers are the typical summer weather in Korea. Arriving on a summer afternoon, a rain shower cools down the raging heat. So, like farmers in a drought urgently waiting for rain, many people look forward eagerly to a refreshing rain shower on a summer afternoon. These two occasions might be the only times when Koreans look forward to the rain.

Rain showers occur when the air covering Korea is unstable. During the summer, such instability is particularly common at midday, so most summer showers come in the afternoon. When the air is unstable, a slight

Cumulus in mountain areas and showers of rain When air moves and hits against a mountain, it ascends and forms fleecy clouds on the windward side of the hill. (Gochang, Jeollabuk-do. June, 2006)

upward draught causes a continuous rising of the air, which leads to the development of thick clouds. So, afternoon skies are often filled with fleecy cumulus clouds.

Under the influence of an unstable North Pacific air mass, there are two factors that contribute to Korea's rain showers. One is the mountains that range from the northeast to the southwest of the peninsula. The other factor is the presence of streams and rivers flowing between those mountains. The North Pacific air moves towards Korea from the southwest. Since the Taebaek range, Korea's backbone, is positioned in the east, the air flowing in from the southwest is forced to ascend further when it reaches them. Passing over the rivers and streams, it gains a lot of moisture. Therefore, thick clouds frequently develop on the western slopes of the Taebaek range, clouds that ultimately release showers of rain.

But mountains aren't entirely necessary for rain showers. In the locale where I work, there are no mountains, yet showers are frequent. On midsummer mornings there is a clear, bright blue sky. But around lunchtime, clouds partly begin to form, and they gather so rapidly that by the time I am walking back to work after lunch, the sky is no longer blue. At around three or four o'clock, cumulus begin to rise in the distance. When the land has been warmed by the heat of the day, the surrounding air is destabilized, and there is a ascending air current. When that current gains power, the clouds thicken and the rains come. With the thick clouds absorbing all the sunlight, it's dark below, even though it is still daytime. In

Rain showers Water drops fluctuate in the clouds and grow in size and finally fall as rain showers.

such conditions, downtown streetlights are turned on in cities.

These showers are usually accompanied by thunder and lightning. Inside a cumulus there are powerful air currents, both ascending and descending. These are dangerous conditions for airplanes. Recently a commercial airplane narrowly averted disaster when flying through such a cloud. People on the ground are also at risk. When there is thunder and lightning, it is dangerous to stand in an open area or to take shelter beneath a tree. Occasionally people are struck by lighting while playing golf.

Sometimes these intense showers include hail. After all, hailstones are just water droplets that have gradually grown bigger after repeated ascent and descent in the clouds. As the temperature changes according to altitude, they are subjected to repeated freezing and melting. Unlike an ice cube continuously frozen in a freezer, hailstones develop in layers marked by lines similar to the growth rings on trees. Hailstones vary in size, with some reportedly as large as ping-pong balls. Hail is deadly to crops, and when hail falls on vegetable beds, it can ruin the whole year's farming.

While it's falling it seems interminable, but the rain subsides in the evening. As if to tease homeward-bound commuters, most showers stop around seven o'clock. Most rain showers last about an hour or so. Although it feels like heavy rainfall, these summer afternoon showers rarely exceed 10 millimeters.

In the middle of a heavy shower, you might phone a friend in the same city and hear that they are standing in sunshine. As the Korean expression goes, "A rain shower divides a cow in two." Just as the rain might wet only one side of cow, darkening its hide unilaterally, so the same city can have neighborhoods drenched in rain at the same time that the sun is beaming down on other parts of town. Since there is so much local difference when it comes to showers, the typical midsummer weather forecast will say, "It will be mostly sunny across the country, with light afternoon showers in some areas." Listeners react sarcastically to such open-ended forecasts, but they are accurate descriptions of Korea's midsummer weather.

09

The high skies of autumn

When the late rainy season is over, and a couple of typhoons have passed by, it is autumn in Korea. As the air in the Siberian plain starts to approach the peninsula, you can finally enjoy the high, clear blue sky, thanks to the torrential rains and midsummer showers that have cleaned the dust from the air. And as the daily temperature range widens, the mountains are covered with beautiful red and yellow autumn leaves.

oreans call autumn the season of high skies and plump horses. But above all, the season's best feature is the golden waves of crops ripening under the hot sun that shines through the clear, high sky. Although I had studied geography, I fully realized the beauty of the golden fields only after seeing a rice field plain in autumn. Since then, I revisit the plains often, especially in autumn, but whenever I have the chance, in all four seasons.

A two-hour drive northeast from Seoul takes you to the vast, open Cheorwon plain. There is also the wide Gimpo plain in the west, much nearer to the capital. But Korea's open rice plains are Mangyeong and Gimje plains in Jeollabuk-do, where the fields stretch endlessly to the distant horizon. Once immersed in the golden autumn scenery there, it is hard to wake from the spell. The Mangyeong and Gimje plains, also

The season's best feature is the golden waves of crops ripening under the hot sun that shines through the clear, high sky. (Cheorwon, Gangwon-do. Sept., 2007)

The golden fields stretch endlessly to the distant horizon at the Gimje plains in Jeollabuk-do. (Oct., 2006)

known as the Honam plains, are far away – about a three-and-a-half-hour drive from Seoul – but I still make the trip two or three times every autumn. I never tire of the plains. The fascination will remain as long as rice is Korea's staple food.

Besides the golden rice fields, many other images come to mind when one thinks of autumn in Korea: the early morning mist that gradually grows denser; the colorful autumn leaves that drop from the trees to

Rows of neatly-cut, bright orange persimmons hung out under the eaves to dry are one of the many images that come to mind when one thinks of autumn in Korea. (Wanju, Jeollabuk-do. Oct. 2005)

decorate the hills and the valleys in the mountain regions; the rows of neatly-cut, bright orange persimmons hung out under the eaves to dry – the dried persimmons that are said to be more frightening than a tiger[1] – and the grass turning yellow on the university campus. All these add to the charm of autumn, and make a person want to set out on a trip to commune with nature. Even for people who aren't much given to travel, the urge is compelling. The Korean countryside is a natural treasure that amply rewards any travel and exploration.

---

[1] Translator's note: The allusion is to an old Korean folktale. In that story, a tiger approaches a small house in the mountains on a cold winter night and overhears a mother trying to make her child stop crying. The mother threatens the child, saying a tiger might come and eat it, but the child just keeps on crying. Then the mother says that if the child doesn't stop crying, she won't give it a dried persimmon to eat. Hearing that, the child immediately stops crying. The tiger is incredulous: humans, it seems, are more afraid of dried persimmons than of a big tiger.

Why does the autumn sky seem 'high'?

As a child, there were many days when I would lay out the straw mat and take an afternoon nap. The mat spread on top of some barley straw cut in the early summer was good enough for bedding. The family would eat dinner on that mat in the summer, too. And when it was time to clear away that straw mat, it was autumn. It is difficult to sleep outdoors because of the autumn dew. Around that time, if one lies on a straw mat during the day and looks up to the sky, it really seems to?have become higher.

Normally, people regard autumn as the period from September to November. But really it starts as soon as the late rainy season is over. At that time, the temperature drops sharply from the midsummer high. The weather is already cool enough to be called autumn. Such pleasant weather starts from late August. The occasional typhoons that pass Korea seem to accelerate the coming of autumn; once they have passed, they are immediately followed by the ideal weather of clear blue skies and cool breezes. But it's only after early September that the high autumn skies make their full appearance.

From midsummer until early September there are so many rainy days that it's hard to speak of high skies. But once the late rainy season is over and a couple of typhoons have passed by, autumn has truly arrived. As the air in the Siberian plain starts to approach Korea, the bright blue sky seems higher. This is because there is little dust in the air and, therefore, less scattering of sunlight in the lower altitudes.

Scattering occurs when the sunlight hits small molecules of air. These molecules disperse the shorter (blue) wavelengths of light more than the longer (red) ones, which is why the sky looks blue to us. Without such scattering, the sky would have no color at all; it would seem black. In fact, past the earth's atmosphere, out in space, it is always dark. That is because there are no air molecules to scatter the light. For the same reason, if you look out the window of an airplane, once you pass a certain altitude, the

In higher altitudes, there is no air to disperse the sunlight. Therefore, if you look out the window of an airplane, once you pass a certain altitude, the sky turns dark. (The skies above Uzbekistan. July, 2006)

sky turns dark. Large dust particles also cause dispersion, but because they are at a lower altitude, the effect is to make the sky look hazy.

Right after the end of the late rainy season is when there is the least amount of dust in the air, thanks to the cleansing effects of the torrential rains and midsummer showers. Consequently, since there isn't much dust to scatter the light at lower altitudes, the scattering effects occur much higher up: the sky seems 'higher', and thanks to the absence of haze, much brighter. By the same process, the sky is clearer and seems 'higher' the day after a rain.

In early autumn the sun's angle is still relatively high, so daytime temperatures continue to be fairly warm. However, as the influence of air from the Siberian plain increases, nighttime temperatures begin to drop sharply. This leads to an increase in the dirunal temperature range. In fact,

In autumn, there is an increase in the diurnal temperature range, that is, the gap between the daytime high and nighttime low temperature, so at dawn when the temperature falls below the dew point, fog often sets in. (Gokseong, Jeollanam-do. Oct. 2006)

along with the time when the northwesterly wind appears in spring, early autumn is the time of year with the greatest diurnal temperature range. At dawn the temperature falls below the dew point, and fog occurences. The air is cold enough for water vapor to condense as dew or fog. Mist is so frequent at this time of year that we call it the herald of autumn. Reports of serious traffic accidents are also frequent: unfortunate consequences of the early morning fog.

Why are autumn leaves so beautiful?

When I first moved to Seoul, I was much impressed by the yellow leaves of the ginkgo trees that lined Sejongno, Seoul's main downtown thoroughfare. Growing up in the countryside, I had never regarded autumn leaves as objects of beauty. Back then, one rarely saw a ginkgo tree

The presence of bright yellow ginkgo trees brings radiance to the streets of Seoul. (Yangje, Seoul. Nov., 2007)

The maple trees on high mountains are the first to change color as the diurnal temperature range increases and the temperature falls in autumn. (Jangseong, Jeollanam-do. Oct. 2005)

on Jejudo Island. But the presence of those ginkgo trees brought real radiance to the streets of Seoul.

The first time I saw mountaintop autumn leaves up close was some ten years after first admiring the ginkgo leaves in Seoul. They were the autumn leaves at Baegyangsa, a temple on Baegyangsan, in Naejangsan National Park. Throughout the country, Baegyangsa is famous for its red maple leaves. Previously I had seen autumn leaves on mountains from a distance, but I had not never been on the actual mountain where I could nearly touch the leaves. It is difficult to express the beauty I experienced back then. I go on field trips as often as most people go to their offices, but it wasn't until ten years later that I returned to Baegyangsa for the autumn leaves. I had visited the nearby area often, but it is never easy to meet the exact timing for the autumn leaves. Seoraksan also has beautiful autumn leaves. Moreover, the maple trees there are the first to change color in South Korea. The autumn leaves start to appear at Seoraksan in early October, and then gradually move down south, colorfully dyeing the trees throughout the country by late October.

The autumn leaves on Jejudo Island's Hallasan fall far short of the beauty of the displays on Seoraksan or Naejangsan. The reason that autumn maple leaves on the mainland are more beautiful than on Jeju is that the former are exposed to a greater range of temperatures during the daily cycle. Along with sufficient sunlight, that diurnal temperature range is the crucial factor in producing beautifully colored leaves. Among the three countries of Northeast Asia, Korea has the most spectacular displays of autumn leaves.

Around the time that the leaves on Seoraksan take on their spectacular

In autumn, mouth-watering red apples are ripe for the picking. (Yeongju, Gyeongsangbuk-do. Oct., 2007)

hues, the fields start to turn into a sea of gold. And come October, those fields are filled with farmers busily gathering the year's harvest. In the neighboring orchards, mouth-watering red apples are ripe for the picking. The harvest season has arrived. Gyeongsangbuk-do which is famous for their apples, are in an area of great diurnal temperature range. Less cloudy in autumn than other regions, the days are warmed by intense sunlight.

Fortunately, Korea's cool, clear autumn weather makes for good rice harvesting. A crop that has dried well commands a good price. But if it happens to rain during the harvest season, the quality of the rice deteriorates. Driving through Gimje during the rice harvest, you can see rice being dried, along with pretty cosmos flowers blooming on the roadside. It is said that the cosmos is not a native Korean flower, but it goes

In autumn, pretty cosmos flowers bloom on the roadside throughout Korea. (Gimje, Jeollabuk-do. Oct. 2007)

well with Korean scenery, waving in the gentle autumn wind. Such a delicate flower would have been ill suited to a country where the winds blow stronger in autumn.

Today, even with a well-dried harvest of rice, a farmer cannot feel content. Koreans eat less rice than they used to, and they import cheaper rice from abroad. Up until about twenty years ago, such a situation was unimaginable. In those days, every bit of usable land was used for rice cultivation. There were even terraced rice-paddies running step-wise up steep, hazardous hills. Today, such rice fields are run more as tourist attractions than actual agricultural enterprises. The Daraengi Maeul (literally, 'rice-paddy footpath village') in Gacheon-ri, Namhae in Gyeongsangnam-do is one such tourist farm.

In the past, every bit of usable land was used for rice cultivation. There were even terraced rice-paddies running step-wise up steep, hazardous hills. Today such rice fields are used to cultivate vegetables and developed as tourist attractions. (Daraengi Maeul, rice-paddy footpath village, Namhae, Gyeongsangnam-do. Sept., 2007)

Today farmers no longer try to convert land to rice paddies. On the contrary, for many years now, people have been planting vegetables on former rice fields. And that trend has been increasing since the Uruguay Round Agreement on Agriculture, in the mid-1990s. Nevertheless, the Korean government is conducting a major land reclamation project, saying that there is a shortage in farming land. Personally, I find this whole situation lamentable. I think we should carefully consider which is more important: preserving Korea's wetlands or reclaiming tidal areas to build new rice fields.

On a rice field where the harvesting has been completed, preparations start for a second crop. There is a time in late autumn when the whole field is covered with smoke. Farmers burn the straw and the remaining rice roots to prepare for a next crop. The ashes left from the fire provide nutrition for

In late autumn, on rice fields where the harvesting has been completed, farmers burn the straw and the remaining rice roots to prepare for a secondary crop. (Gimje, Jeollabuk-do. Oct., 2006)

whatever crop is to be planted next, and the fire also helps to exterminate any vermin left in the straw. But the hazy smoke from these fires lowers visibility in the region, a hindrance to planes trying to take off or land at nearby airports. In Korea, barley was the traditional second crop that farmers planted in the fields after the rice was harvested. But today the double-cropping system includes other items. Garlic has become an important secondary crop, and in the Gimje plain, more farmers now plant potatoes than barely.

Towards late autumn, frost starts to form out in the fields. And when the first snow falls, it is winter. In any region, people start to prepare for winter upon the year's first snow. And whenever that day happens to fall in the calendar, winter has arrived. Already trees have shed all their leaves, and the fields are completely bare. In recent years, the first snow has tended to

Around the time of the last harvesting in late autumn, the first frost sets in. At the bottom half in the photo is a field of beans, which are generally picked after the first frost. (Sunchang, Jeollabuk-do. Oct., 2007)

come relatively late.

The effects of the first snow extend to our emotions. Young people who have been feeling a bit depressed brighten up as soon as they hear news of the first snow. Many people make special, difficult promises for this first snow-day. I have also made such promises, but failed to keep them. Friends I hadn't seen in a long time reproached me about those broken first-snow promises, but somehow I didn't feel too guilty. Nowadays, stores hold many first-snow sales and events. No wonder young people look forward so eagerly to that day. In any case, the first snow can feel like a breath of fresh air, a balm to the gloomy mood of late autumn. That is how the weather can energize our everyday lives.

Part 3

Temperature, Rainfall, Wind, Fog, and Frost

10

Temperature

There is a fairly wide regional temperature variation in Korea. Aside from the predictable difference between northern and southern regions, there is also marked difference between east and west. The reasons for this considerable variation are closely related to mountainous effects and to the seas that surround the peninsula along three coastlines.

Temperature is important in determining the distribution of vegetations and croups products, and also in defining the limits of human habitation. From early on, temperature was an important element in the classification of climatic zones. Temperature itself is the main factor in the regional distribution of vegetations and crops, but in terms of the everyday lives of human inhabitants, the key element is wind-chill.

On Jejudo Island, Korea's southernmost region, orchards of tangerine orange trees were first planted in the 1960s. The supply of Japanese cedars as windbreak forests greatly encouraged farmers to plant more orchards. At first, tangerine were grown only in and around Seogwipo, in the warmer southern part of the island. But thanks to the windbreak forests, orchards spread across the island. However, the tall trees of the windbreak forests changed the scenery of the villages. Trees in a mature windbreak forest reach two or three times the average adult's height, so villages lost their compact, human-scaled feeling, and were lost to sight from the outside.

Forests and stonewalls are natural windbreaks on Jejudo Island. (Jeju, Jeju-do. Jan., 2007)

And for the villagers, it became difficult to enjoy a view of outside vistas.

First, Jeju's farmers would plant tangerine orange trees in any place that seemed warm. But that turned out to be a misguided strategy. In areas where the wind is strong, the fact that a place feels warm doesn't necessarily mean that the temperature there is high. In fact people feel warm in places where the nighttime temperature actually drops further than it does elsewhere. A place that is lower in altitude than its surroundings is warm during the day, but that is where cold air gathers at night. In such places, the tangerine trees were damaged by cold weather.

During the severe cold wave in January 1990, subfreezing weather lasted for nearly a week even on Jejudo Island. Under such conditions, tangerine trees can't last long, so many of them froze to death in certain orchards around the island. Most of the trees that died were in unventilated greenhouses. Wind heightens the chilling effect of low temperatures on

The Jeju Provincial Government encouraged farmers to cut down their tangerine trees and to replace them with black raspberry bushes that are enjoying a new popularity. (Jeju, Jeju-do. July, 2008)

human beings. But by cycling the air, wind actually prevents temperatures from dropping sharply. So, the cold-temperature damage to trees occurred in places that, to people, felt relatively warm.

The orchards have withstood the fierce sea wind, but as commercial enterprises, they are undergoing hardship as the Korean market is gradually opening to imported fruits. Once the tangerine harvest is over, banners are hung to encourage farmers to thin out their orchards. Subsidies are given to orchards that have completely cut down their tangerine trees. They are being replaced by black raspberries (*bokbunja*) which are enjoying a new popularity nationwide. But just as the price of tangerine oranges have fallen due to overproduction, one fears that *bokbunja* might also become a glut on the market one day.

Why is Korea colder than other countries at a similar latitude?

South Korea is located at between 33 and 38 degrees north, but its

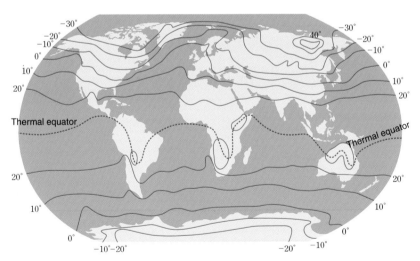

A map of the worldwide surface air temperatures in January Korea's temperatures are generally lower than those in other countries at similar latitudes.

temperatures are generally lower than those in other countries at similar latitudes. That would include such southern European countries as Portugal, Spain, Italy, Greece, and Turkey. The annual mean temperature of those countries is nearly 5°C higher than that of Korea. Even in Japan, though it is not easy to make a simple countrywide comparison because the country extends much further, north-to-south, than Korea, the cities are much warmer than Korean cities at similar latitudes.

The reason for Korea's lower annual mean temperature is because of its exceptionally cold winters. Korea's winter temperatures are about 10°C lower than those of the other countries at similar latitudes. Italy's capital, Rome, is located about 41 degrees north, and its mean temperature for January is 8.8°C; however, in Seoul, at just over 37 degrees north, the mean temperature in January is minus 3°C. In winter, Southern Europe is

The fields in the middle region of Korea One cannot plant crops outdoors in winter from the middle region up to the northern provinces of the peninsula because of low temperatures. (Hongcheon, Gangwon-do. Feb., 2007)

affected by the warming influence of the Atlantic Ocean and Mediterranean Sea, as well as by the mid-latitude cyclones that pass through frequently, bringing rains that prevent a sharp fall in temperature.

Because its winters are so much colder, Korea's natural scenery is quite different from that of southern Europe. In Korea, except for the southern coastal area and Jejudo Island, it is very rare for farmers to plant crops outdoors during the winter. What farming there is would be vegetables grown in greenhouses, or the snow-covered fields of second-cropped barley. In the warmer southern coastal area and on Jejudo Island, winter cabbages or garlic are cultivated on the outdoor fields. Winter cabbages were first produced in Haenam, Jeollanam-do, and as more farms plant them, they are being grown further and further north. Winter garlic cultivation had also been limited to the southern coastal area, but today, garlic is being

Houses in southern Europe, where it's hot in the summer, have small windows to minimize the amount of sunlight entering the house, with exterior window shutters ready for further shielding from the sun. (Evora, Portugal. Aug., 2008)

planted along the west coast as far north as Seosan, Chungcheongnam-do. In Southern Europe, on the other hand, such grains as wheat are cultivated during the winter. Not only is the temperature warm in winter, but it also rains frequently, so conditions are good for growing grains. This is a major difference from East Asia, where grain farming is mostly restricted to the summer.

The difference in temperature has also affected the design of houses. Korean houses are built primarily as shelter against the winter cold. This becomes more obvious the further north you go on the Korean peninsula; the effects of the Siberian air mass increase as one moves north. In southern Europe, on the other hand, the main consideration in building houses has been the need to keep comfortable during the hot summer season. Thus, windows are kept small to minimize the amount of sunlight entering the house, with exterior wooden shutters ready for further shielding from the sun. Some southern European houses even have a cool, naturally ventilated inner courtyard.

What accounts for Korea's extreme regional tempertuare differences?

The area of Korea is fairly small, yet there are great temperature differences between different regions of the country. While the annual mean temperature of Junggangjin, in Pyeonganbuk-do, or Samsu and Gapsan, in Hamgyeongnam-do, is 0°C, in Seogwipo, on Jejudo Island, it is nearly 16°C. That northern areas should be much colder than southern regions of the peninsula is to be expected, but there are also large temperature differences along the east-west dimension, between areas at the same latitude. The main factor in these regional differences is the combined effects of the mountains and seas.

When two regions at essentiallly the same latitude show a significant difference in temperature, it is generally due to the influence of mountains.

A distribution chart of the annual mean surface air temperatures of Korea (Unit: °C) The isothermal line between the temperature zones of Korea corresponds almost exactly to the extended ridge of the Taebaek Mountains.

A glance at a map showing the distribution of annual mean temperatures around Korea will confirm the role that mountains play. This is especially clear in the case of the Baekdu Daegan Mountain Range, a long chain of many ridges that starts from Baekdusan, in Hamyeongbuk-do, runs south along the Hamgyeong Mountains (Hamgyeongsanmaek), to the Taebaek Mountains, turns west towards the Sobaek Mountains (Sobaeksanmaek), ending finally at Jirisan. The pattern of isothermal line in Korea corresponds almost exactly to this extended ridge of mountain.

Also, since it is a peninsula, the pattern is influenced by the seas that border it on three sides. Moreover, there is a huge difference in size between the East Sea and the Yellow Sea, so they affect temperatures on the peninsula in different ways. The Yellow Sea is relatively shallow: a continental shelf with an average depth of 128 meters. The East Sea, on the other hand, is much deeper, extending down to 2,000 meters in places. Given the thermal inertia of water, deeper bodies of water are subject to much less temperature variation than shallow ones. Thus, while there is almost no seasonal change in the temperature of the East Sea, the Yellow Sea shows significant temperature variations according to the time of year. And these variations affect the climate on the land nearby.

So, both the mountains and the seas figure in the differences in termperature between the eastern and western regions of the Korean peninsula. Temperatures are higher in Gangwon-do, on the east coast, than in areas on the west coast, largely because the Taebaek Mountains block the cold winter northwesterly wind. Another factor raising temperures on the east coast is that the deeper waters of the East Sea are less subject to seasonal variation than the shallower Yellow Sea waters off the west coast, which become relatively cold in winter. Due to the combined effects of mountains and seas, the mean temperature for the month of January is about 3°C higher in Gangneung than in cities and counties on the west coast, such as Incheon or Ganghwa. But in the summer, the temperatures are quite similar in the two regions.

A 3°C difference in the January mean temperatures is not insignificant. Since it's the coldest month of the year, that small difference can crucially determine whether and where frost and freezing conditions occur. The temperature difference between the two sides of the peninsula is reflected in the scenery. In the east-coast city of Gangneung, persimmon trees line the urban streets, and there are stands of bamboo in residential areas. Persimmon and bamboo trees can only grow where winters are warm. In Ganghwa, on the west coast at a similar latitude, bamboo and persimmon trees are not to be found. However, the sight of orange ripening persimmons has recently become common further inland, in downtown Seoul.

The sight of orange ripening persimmons has recently become common even in Seoul. (Seoul Arts Center. Nov., 2007)

The temperature difference between the northern and the southern regions is also significant, since the north-south extent of the peninsula is fairly long. Between the northernmost point (43 degrees, 05 minutes north: Yupo-myeon, Onseong-gun, Hamgyeongbuk-do) and the southernmost (33 degrees 06 minutes north: south of Marado, Daejeong-eup, Jejudo), there is a latitude difference of 10 degrees. That's a distance of over 1,000 km.

In addition, the fact that the northern tip is connected with the Eurasian Continent, and the southern end looks onto the vast Pacific also

In the winter, in the Cheorwon plain, one cannot see any green at all. (Cheorwon, Gangwon-do. Feb., 2008)

Garlic cultivated on the outdoor fields In the southern Namhae coast, all the fields are green even in winter. (Haenam, Jeollanam-do. Feb., 2008)

contributes greatly to the difference in temperature between the northern and southern parts of the peninsula. It also explains why the temperature difference is most pronounced during the winter. In the summer, in August, the place with the lowest temperature in Korea is Samjiyeon (in Yanggang-do), while the warmest place is Seogwipo, in the south of Jejudo Island. But the difference between the two is only 10.7°C. However, in the winter, the difference is nearly 25°C. It is this enormous difference in winter that accounts for most of the difference in annual mean temperature. Comparing Cheorwon, the northernmost region of South Korea, with the southern coast, there is hardly any difference in their natural scenery in the summer season. Both present a verdant scene of overgrown greenery. But come winter, the situation changes completely. In the winter, in the Cheorwon plain, one cannot see any green at all. Only migrating birds fly above the desolate earth. But in the southern coast or on Jejudo Island, all the fields are green even in winter. Most of the fields can produce winter crops.

One winter, I went on a field trip to the southern coastal area of Jeollanam-do with some soon-to-be second-year university students from the geography department. For many of them it was their first time out of the middle region of Korea. As we entered the Jeolla region, they looked out the window with wonder in their eyes. At first, they thought they were mistaken, but indeed what they were seeing, for the first time in their lives, were green fields in winter. Even if they had visited the region before, it would have been in the summer, for a trip to the beach, or for a school trip in the spring or autumn. And ever since, I have described their amazement to students who are accustomed to travelling across the country several times a year for visits back home, to remind them of the important determining effects climate has on landscapes.

The temperature difference between the northern and southern Provinces has also had an effect on the traditional design of the houses. The houses of

The architecture of houses in the east-coast area of Gangwon-do show traces of a local historical similarity to the houses in Hamgyeongbuk-do; the kitchen is directly connected to the wooden-floor sitting room (*maru*). (Wanggok village, Goseong, Gangwon-do. Jan., 2000)

A farmer's house in the southern region of Korea A narrow wooden hallway (*toenmaru*) along the exterior wall of the house remain cool in summer, and helps one overcome the sweltering heat. The wooden floor, being slightly elevated, the ground heat is carried away in the breeze. (Jindo, Jeollanam-do. March, 2008)

Hamgyeong-do, where the winter temperature is low, were built so that the heat from the kitchen would spread evenly inside the house. And to make use of that heat, a living room was built between the main kitchen area and the other rooms of the house. But there was no wall or door between the kitchen and living room.

In the east-coast area of Gangwon-do, which historians say was closely associated with the Goguryeo Kingdom during the Three Kingdoms Era (37 BCE- 668 CE), you can see traces of a local historical similarity to the houses in Hamgyeongbuk-do. At Wanggok village in Goseong, Gangwon-do, a preserved historical folk village with actual residents living year round, there is no door between the kitchen and the wooden-floor sitting room (*maru*). According to the inhabitants, the living room had the same role as the kitchenette, and it is warm there, thanks to the heat from the kitchen.

In contrast, the houses in the southern regions of the peninsula traditionally had a big, open wooden-floor sitting room, called *daecheong maru*. Because of the hot, humid summers, it was important for people to prepare for summer discomforts, as much as for the winter cold. In the traditional houses of ordinary farmers, there was a narrow wooden hallway (*toenmaru*) along the exterior wall of the house, especially outside the door of each room. The wooden floors remained cool, and helped one overcome the sweltering heat.

Where are the hot and the cold places in Korea?

Daegu is known as the hottest place in Korea. That's because it is located in an inland basin, so it receives extra heat from the hot, dry down-slope wind that occurs in the lee of mountain ranges. But this situation is hardly unique to Daegu: nearly all of Korea's are surrounded by mountains. With the exception of coastal ports, settlements in Korea were almost always developed in the fields beside a river, generally surrounded by mountains.

The mean temperature of major cities in August, and the number of hot days the daily high exceeds 30°C

Observation point	Mean temperature (°C)	Number of hot days
Daegu	26.1	55.3
Sancheong	25.0	46.6
Yeongcheon	25.1	46.8
Uiseong	24.7	49.5
Jeonju	26.1	48.5
Masan	26.6	37.6
Gwangju	26.1	44.9
Jeju	26.5	30.5
Seogwipo	26.6	24.7

(Source: Korea Meteorological Administration, 2001)

In August, the mean temperature of Daegu is 26.1°C, and with its surrounding areas, it is considered the hottest place in Korea. Compared to other cities, like Jeju (26.5°C), Jeonju (26.1°C), Masan (26.6°C), and Gwangju (26.1°C), Daegu is not that much hotter. It would hardly seem to deserve its reputation as Korea's hottest place. But if you count the number of days when the daily high exceeds 30°C, you get a better sense of how hot it is in Daegu. Averaging 55.3 hot days a year, Daegu leaves the rest of the country behind; Uiseong, in Gyeongsangbuk-do, is a distant second with 49.5 hot days anually. Seogwipo, which has the highest annual mean temperature, records fewer than 25 days a year when the tempearture exceeds 30°C. Recent meterological recoreds indicate that Hapcheon and Miryang, in Gyeongsangnam-do, are also high-temperature areas.

Just as there are the hottest places in summer, there are also the coldest places in winter. However, the inhabitants of such places do not welcome the designation. They fear that if their hometown becomes known for its winter cold, outsiders will shun the region. This shows how greatly the winter cold affects our lives.

If Daegu is a hot place in the summer because it is located in a basin, it could be expected to be cold in the winter for the exact same reason. Basins tend to become cold as cold air descends from the surrounding mountains at night. Yet it is not that cold in Daegu; the mean temperature in Jaunary is 0.2°C. Apart from the mountain areas, the place in South Korea with the lowest mean temperature in January is Hongcheon, in Gangwon-do, at minus 5.6°C. Hongcheon is a small town in a mountain basin. Other places that are known for their winter cold include Cheorwon, and Wonju in Gangwon-do, Yangpyeong in Gyeonggi-do. It is also relatively cold in Uiseong and Bonghwa, in the inland region of Gyeongsangbuk-do.

Yangpyeong also gained some notoriety for it winter cold. In January 1981, the temperature there fell to minus 30°C, cold enough to break a bottle of Korean distilled alcohol (*soju*) that was displayed at a store. Thanks to the national media's reporting of that incident, Yangpyeong gained a reputation for cold. Indeed, the area is surrounded by high mountains, and moreover, when the nearby Hangang freezes, the winter sun is reflected and little heat is absorbed. The combination of these factors accounts for Yangpeyong's deep-freeze winters.

11

Most of Korea's rain comes in the summer

Korea's annual precipitation is considerably higher than that of most countries. Yet every year, Korean farmers worry about a shortage of water for their crops. That is because most of the rain falls during summer. The summer rainy season lasts more than a month, during which nearly every other day sees heavy rains. It is followed by a period of frequent showers, under the influence of the hot, humid North Pacific air mass. During that period, a great deal of rain falls whenever a extratropical cyclone passes across the peninsula.

Nowadays the public supply of tap water reaches over 80 percent of South Korean households, so most families hardly worry about a shortage of drinking water. But until the 1980s, only 55 percent of households had tap water, so the sight of people carrying buckets from public wells or nearby ponds was common.

In the mid-highland villages of Jejudo Island, people lived on water obtained from a small pond. That pond was home to tadpoles, with snakes paying occasional visits. But once autumn was over and winter had arrived, even that small amount of water would dry up to reveal the bottom of the pond. Then the search for water would make the village women walk for two or three hours up Hallasan in order to reach another pond. Of course they had to carry a vessel for the water. But most of the time the contents of that pond was half water, half mud. If it happened to snow, the villagers would melt pots of snow on a fire and used it as drinking water. Needless

In the mid-highland villages of Jejudo Island, people lived on water obtained from a small pond by the side of a stream. (Seogwipo, Jeju-do. Jan., 2008)

to say, the water from the melted snow was full of dirt as well.

In the mid 1990s, I went on a midsummer field trip with some students to Chujado Islands, northwest of Jeju. Although it was the time of year when water is overflowing in most parts of Korea, the elementary school where we wanted to stay made it clear that they could provide us accommodation, but not any drinking water. There was no other choice than for each person in our group to carry a big plastic 18 liter container of water to the island. Initially I found it hard to believe that the water situation was really so dire, but once on the islands, I immediately grasped how scarce and precious water was there. All available space at every house was occupied by some sort of container to catch and save rainwater. Whenever it rained, the residents of the island struggled to save every precious drop.

Some five years after the trip to the Chujado, I visited Ulleungdo Island

All available space at every house is occupied by some sort of container to save rainwater on Chujado where fresh water is a scarce resource. (Chujado, Jeju-do. April, 2008)

for the first time. Remembering the experience on Chujado, I carried plenty of water onto the ferry. But the first word from an employee of the Ulleung county office was that the people on the island were very generous about offering water. He told me that there was plenty of water everywhere on the island, so I could drink as much as I wanted to without any worry. And indeed all the villages had an overflowing supply of water. Though both Chujado and Ulleungdo are islands, their situations are totally different as regards water.

Unlike other regions in Korea, there are many cloudy days on Ulleungdo Island, and it rains evenly throughout the year. In most other areas, it rains mainly in the summer, but on Ulleung, the amount of precipitation is similar in each season. Actually, the amount is somewhat higher in the winter, due to the large snowfall. Since precipitation is evenly distributed throughout the year, the plants and the people living on the island came to learn how to use rainwater effectively for their needs.

Why are there so many floods and droughts in Korea?

Korea's annual precipitation is about 1,300 mm considerably higher than the global average of 800 mm. But despite that abundant rainfall, shortage of water for their crops is a perennial worry for Korean farmers. The reason, of course, is that most of the rainfall occurs during the summer. So, both heavy rain and draught give cause for concern. With the exception of Ulleungdo Island, in nearly all regions nearly half of the total annual precipitation falls during the three summer months. In extreme cases, rain equal to an entire monthly mean can fall in a single day.

On August 31, 2002, during Typhoon Rusa, the Gangneung region recorded 883.2 mm of rainfall. That is the greatest single-day rainfall in the history of Korean meteorological observation, and is close to two thirds of the average annual precipitation for the Gangneung region. This torrential downpour led to massive flooding. Indeed, the unprecedented flood damage

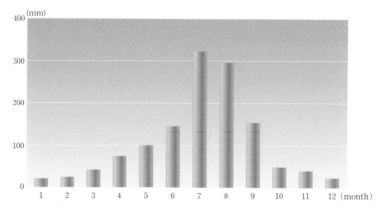

The monthly rainfall distribution chart of Hongcheon, Gangwon-do Like most parts of Korea, most of the rainfall occurs during the summer in Hongcheon.

Seven months after a flood inundated the area, a damaged road is still to be repaired. (Gangneung, Gangwon-do. March, 2003)

put the city's future in risk. When precipitation over 80 mm is predicted, the Korea Meteorological Administration issues a heavy rain advisory; when the forecast amount exceeds 150 mm, an official alert is broadcast. So you can imagine what degree of alarm would be justified for nearly 900 mm of rain in a single day. It is doubtful that we'll see that record broken in our lifetime.

Although the pattern varies slightly by region, overall, the dry and the rainy seasons are clearly distinguished in Korea. The wet season extends from the start of the rainy season in late June through the beginning of autumn in mid-September. The rest of the year – autumn, winter, and spring – can be called the dry season.

In midsummer, the air is unstable, so there are frequent showers. In addition, torrential downpours occur when extratropical cyclone pass over the peninsula. Every August, without fail, there are news reports of campers being stranded in mountain valleys, cut off from surrounding settlements by swollen streams. These streams are typically swollen by flash flooding due to the torrential rains brought on by passing extratropical cyclone.

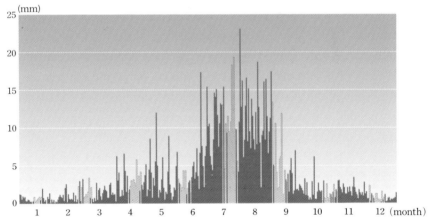

The monthly average rainfall in the Seoul region There is a clear distinction between the dry and the rainy seasons in Korea.

Rain is also frequent during the period from late August to mid-September; therefore, it is known as the "late rainy season." During this time, when the polar front that had moved up north starts to move gradually down south, Korea and the surrounding regions are subject to typhoons. Thus, climatic conditions in Korea are conducive to rain from late June through the middle of September.

The fact that it is the wet season doesn't mean that it rains continuously If a polar front is slow to yield precipitation or moves beyond the boundaries of Korea, the weather situation can change dramatically. When rains do not materialize during the rainy season, it is called a "dry rainy season," and the whole country suffers from a drought. At such times, people even pray for the devastating typhoons, if only for the sake of the rain they bring. News reporters have coined the term 'filial son typhoons' in reference to those beneficial effects.

The heavy rainfalls in the late rainy season cause flooding. That is because the ground is already saturated, so when more water comes from torrential rain, it cannot be absorbed. Nor is the danger limited to the water-saturated ground. At this time of year, Korea's huge dams are also almost filled to capacity. Hence, it is a very stressful time for the people responsible for the water management facilities. It isn't easy to make the right decision – whether or not to open the dams' floodgates – because one never knows if that day's rainfall will end up being the last significant precipitation of the season. Furthermore, in late summer, even a moderately heavy rainfall can destroy a crop in its final ripening stage, ruining the year's harvest. Clearly heavy rainfall is an unwelcome guest in late summer. It is nearly impossible to recover the crop from a rice field that has been inundated at that time of year. On the other hand, if there is no rain during this period, there could be a severe water shortage in the following winter and spring. In the worst case, when the rice-planting season rolls around in late April or early May, there might not be enough water to

Rice fields inundated by a heavy rainfall in the late rainy season. (Gyeyang, Incheon. Sept., 1984. Kwon Hyok-jae)

If it doesn't rain in late August or early September, there could be a severe water shortage from autumn through the following winter and spring. (Okjeongho, Jeollabuk-do. Nov., 2008)

During the dry season, many small streams in Korea have virtually no water, revealing a dry riverbed. (Ssangcheon, Sokcho. Jan., 2008)

irrigate the new crop.

The twin dangers of flooding and draught are exacerbated by the topographical features of Korea's rivers. To begin with, rivers on the peninsula typically have a small drainage area. So, immediately after a heavy rain, the water level rises drastically. For the same reason, the water rapidly flows downstream once the rain stops. Thus, floods are frequent, but of short duration. Furthermore, since the water flows so quickly, during the dry season, many Korean rivers have virtually no water, revealing a dry riverbed.

Another factor that encourages flooding is that most Korean rivers flow from east to west, while, with the exception of the backbone Taebaek range, most mountain ranges stretch from the northeast to the southwest of the peninsula. As a result, most valleys stand open towards the southwest, so when the southwest air current blows in the summer, conditions are ripe for heavy rains in the valleys.

How Koreans have handled the perennial problems of flooding and drought

Throughout their history, Koreans have faced the problems of floods and droughts on a yearly basis. So, they have had to be prepared for those weather conditions. The fact that Koreans invented a rain gauge in the fifteenth century and have kept records of precipitation for centuries shows that the unpredictability of the rainfall has long been a crucial issue.

When a spring drought becomes prolonged, the year's crop is at risk. Sometimes, when an entire spring month goes by without precipitation, farmers urgently await the rain as they watch their rice fields dry up.

Right before the rainy season, it hardly rains for weeks. But on top of this dry spell, if the heavy rains are delayed in arriving, the early summer heat will dry up the fields even more. In the days before modern irrigation, when rice farming depended solely on rain, when rice-farming was the people's main occupation, and rice their main food, the timing of the onset of the rainy season was a matter of concern to the whole country.

In the central region of Korea, the bottom of the rice paddies where rice plants have been transplanted start to crack, and in the southern region, where it is warm enough to farm two crops a year, farmers cannot start the spring planting until the rice paddies are filled with water. Even now, with a modern irrigation system, farmers still become anxious if the rainy season is late in arriving.

From the old days, Koreans have always put great importance on the timing of the season because if a farmer missed the right time, the year's harvest could be ruined. If a dry spell forced farmers to miss the proper time for rice-planting, they would quickly sow a replacement crop. These crops, such as buckwheat, are hardy plants that can grow to maturity

Nowadays buckwheat is valued less as a hardy replacement crop than as a health food for dieters and fitness-conscious consumers. Buckwheat flowers are an important element in the local festival of Bongpyeong. (Pyeongchang, Gangwon-do. Sept., 2006)

within a short period. However, nowadays buckwheat is valued less as a hardy replacement crop than as a health food for dieters and fitness-conscious consumers. Meanwhile, in the town of Bongpyeong, in Pyeongchang-gun, Gangwon-do, buckwheat flowers are an important element in the local festival, thanks to a much-read novel, *When the Buckwheat Blooms* (published in 1936), Yi Hyo-sok.

Throughout history, Koreans have regarded water as a very precious resource, and from centuries ago they built dikes and reservoirs to manage it. The dikes were built as water-conservation facilities along narrow streams at places where it was difficult to build a reservoir. Even now, especially in the mountain areas of Gyeongsangbuk-do and Gangwon-do,

Dikes in the mountain areas In the mountain areas, one sees many small pools where water is stored for later release along small channels to nearby rice paddies (left in the photo). The stairs-like facilities in the center are the waterway for fish travel to the upper or lower reaches of the stream. (Samcheok, Gangwon-do. Feb., 2006)

one sees many small pools where water is stored for later release along small channels to nearby rice paddies. These small dikes that are built in the small valleys clearly show how highly Koreans treasure water, and how they have devised ways to use the natural resource effectively.

Reservoirs are an important water management facility even today. Since the 1970s, several large multi-purpose dams were built throughout the country, but reservoirs still serve an important role. Historically, there were such reservoirs as Uirimji in Jecheon, Byeokgolje in Gimje, and Gonggeomji in Sangju. Among them, all that remains are some traces of the Byeokgolje reservoir. In autumn, the Horizon Festival is held on the empty grounds in front of Byeokgolje, celebrating the good harvest. Gimje is the only place in Korea where one can see all the way to the horizon without any intervening mountains.

The restored Byeokgolje dike (on the right) and the main irrigation channel that distributes water to the rice fields in the Gimje plains. (Gimje, Jeollabuk-do. Aug., 2006)

Today reservoirs provide more than half of the irrigation that rice paddies require. The Daeari reservoir in the upper reaches of the Mangyeonggang is one example of a fairly large reservoir designed for agricultural purposes. The Daeari reservoir was built, simultaneously with a project to artificially straighten the course of the Mangyeong. That is, in modern terms, it could be seen as part of the development of the Mangyeong plain. However, because the water of the Daeari reservoir was not sufficient for the whole Mangyeonggang basin, two more reservoirs were built further upstream: the Dongsang and the Gyeongcheon. And when the combined waters of those three reservoirs prove inadequate, water is even brought in from the Yongdam Dam, in the upper waters of the Geumgang.

The situation is the same for the Dongjingang basin, where water was

Daeari reservoir built on the upper reaches of the Mangyeong back up water so it can be diverted to irrigation channels. (Wanju, Jeollabuk-do. Oct., 2006)

The Chilbo Hydraulic Power Plant draws water from the Seomjingang Dam to the Dongjingang. (In the photo, the three white pipelines on the mountain contain the water resource from the Seomjingang on the other side of the hill. The water flowing on the bottom left is the Dongjingang.) (Jeongeup, Jeollabuk-do. Oct. 2006)

also in short supply. The Seomjingang Dam was constructed, which diverted water towards Chilbo, to cascade down into the Dongjingang. In its drop from the Seomjingang to the Dongjingang, the water turns the hydroelectric generator turbines of the Chilbo Hydraulic Power Plant. This method of generating hydroelectirc power by drawing water from one river into another is called a river diversion power generation.

A shower wets only one side of the cow

Among Korean proverbs about the weather, there is this expression: "A shower wets only one side of the cow." It's a colorful, if somewhat exaggerated metaphor for the extremely local quality of weather in Korea, where heavy rain can be falling in one place, while in an immediately neighboring area the sun shines brightly. The same phenomenon of extreme local differences extends to regional patterns of annual precipitation. Korea has rainy areas, where the annual precipitation is close to 2,000 mm. But in other parts of the same country, the amount is less than 1,000 mm. The country's greatest annual precipitation, 1900 mm, is recorded on the southeastern slopes of Hallasan, on Jejudo. And almost as much precipitation falls on the southern coast of the southern coastal area. The annual average precipitation exceeds 1,500 mm in the areas around Jirisan, at Daegwallyeong Pass in the Taebaek Mountains, and in the areas of northern Gangwon-do along the east coast.

Such extreme amounts of rainfall are due to the surrounding topography. In the case of Jejudo Island, the determining feature is Hallasan. Thanks to that high and massive mountain in the middle of the island, when the southeasterly wind blows, there is a lot of precipitation on the southeastern slopes, and when the northwesterly wind blows, there is a lot of rainfall on the northwestern side.

On a summer's day, if two vacationers who have been camping on two different sides of Jeju meet for dinner, it is quite likely that they will describe

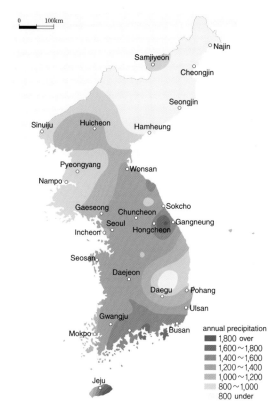

The distribution chart of
**The distribution chart of
Korea's annual precipitation
(Korea Meteorological
Administration)**
The position of mountains
greatly affect rainfall in Korea,
and it rains more on the
windward side, and less on the
leeward side of a mountain.

annual precipitation
- 1,800 over
- 1,600~1,800
- 1,400~1,600
- 1,200~1,400
- 1,000~1,200
- 800~1,000
- 800 under

two completely opposite experiences of beach weather. A vacationer who heads for the beaches on the east side of the island on a day when the wind is blowing from the east will spend a day in overcast, drizzly weather. Meanwhile, another vacationer who headed to a beach on the west side will be enjoying all the pleasures of a bright, sunny summer day.

In addition, the areas that have a lot of rainfall are usually where the southwesterly wind, the periodic wind of summer, blows against mountains to form a ascending air current. This occurs in such places as the environs of Jirisan, and the river basins of the Hangang and the Geumgang.

The influence of topography on the distribution of rainfall Clouds are formed around the top of Hallasan. Due to these clouds, rain showers fall, but it will be bright and sunny on the other side of the mountain. (Jeju, Jeju-do. Aug., 2007)

Precipitation is quite heavy at Daegwallyeong Pass because it is a high mountain area. Located on a ridge of the Taebaek Mountains, it is likely to receive rain no matter which direction the wind is blowing from. If the wind blows from the west, the area is affected by a ascending air current, but conditions are highly conducive to heavy rainfall when the northeasterly wind blows in the winter, and between late spring and summer. It is more difficult to compare the volume of snow and rain that falls on the mountains and on low land, and to give an explanation for the differences. But based on experience, a person who has hiked up a high mountain will easily recognize that mountain areas experience more precipitation.

The inland areas of Gyeongsangbuk-do and the west coast region are places with little rainfall. Among them, Uiseong, Andong, and Gumi, in inner Gyeongsangbuk-do, and the city of Daegu, record only about 1,000

Salt ponds Many salt-evaporation ponds were established along the west coast, where rain is scarce. (Buan, Jeollabuk-do. Oct., 2006)

mm annual precipitaion. Uiseong, with 972.1 mm, has the lowest annual rainfall in South Korea. But for the entire peninsula, the area of lowest annual rainfall is the Gaema Plateau, in North Korea with about 700 mm.

It is difficult for ascending air currents to develop in flatland areas, such as the lower reaches of the Daedonggang river (in North Korea), the shore of Gyeonggiman bay, and the west coast of Jeollanam-do. The fields of the South Jeolla west coast, like Muan and Hampyeong, are dotted with facilities for catching and storing water. The sight of those catchments reminds us of the vital importance of water provision.

In parts of Korea with little rainfall, there have developed industries that actually make use of the dryness. Thus, many salt-evaporation ponds were established along the west coast, where rain is scarce. The production of salt requires ready access to seawater, of course, but more important is the

A pear orchard In places where it doesn't rain much, orchards developed because the fruits are exposed to more hours of sunlight, which result in a higher sugar content. Naju, famous for its Korean pears, is also close to the west coast and has relatively low rainfall. (Naju, Jeollanam-do. April, 2007)

relative lack of rain. So, areas where salt ponds were established are places where a large daily tidal range exposes extensive mud flats; where there are long periods of daily exposure to bright sunshine; and where rainfall is rare. Thus Gwangnyangman bay, in the estuary of the Daedonggang, which is said to have been Korea's first salt bed, is second only to the Gaema Plateau among parts of the peninsula receiving the least precipitation. The average annual rainfall of Gwangnyangman is less than 900 mm. The other salt-pond areas, including Gyeonggiman, Gomsoman, and the area of Sinan, all have relatively little rainfall. Some places that were famous for producing salt through evaporation, such as Gunja and Sorae in Gyeonggiman, have been turned into a huge industrial complex and residential area as part of a land reclamation project.

The fact that the area near Daegu, in rural Gyeongsangbuk-do, gained fame for growing apples has a lot to do with the relatively low rainfall there. And the reason that Naju, the former capital of Jeolla-do, is famous for its round Korean pears can also be attributed to the relative scarcity of rain in that area. Since it doesn't rain much, the fruits are exposed to more hours of sunlight, resulting in a high sugar content that consumers appreciate. Hence orchards are mainly found in places with little rainfall. Nowadays, one finds orchards in most parts of South Korea, apart from areas with particularly high rainfall. A great variety of fruits are now grown as well.

Where does the snow fall in Korea?

Any Korean who has studied geography in high school will immediately think of Ulleungdo Island when they hear the word "snow". Ulleung has the greatest amount of rainfall in the winter in Korea, and it is also just about the snowiest place as well, with an annual average snowfall of 232.8 cm. That figure is second only to Daegwallyeong, which records 258.8 cm. But taking into account that Daegwallyeong is on a highland, Ulleungdo Island may deserve first place for the greatest amount of snowfall as it receives snow on an average of 57.8 days, nearly two months, per year. This is a particularly remarkable statistic considering that Ulleung has a slightly higher temperature in winter than other places at the same latitude.

It snows on Ulleung mainly when the cold, northwesterly winter monsoon is blowing. The air over Korea at this time is extremely cold. But in the East Sea, the sea surface temperature does not drop sharply because the East Korean Warm Current is flowing. The interaction of cold air and warm sea surface produces many clouds. When these clouds hit the bell-shaped Ulleungdo Island, they rise rapidly, growing thicker as they ascend the mountain, until they release their moisture as snow. Thus the amount of snowfall is greatest in the mountains, specifically in the area north of

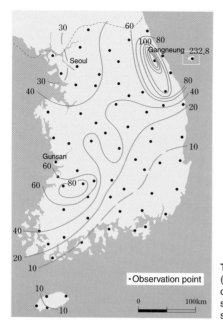

The distribution chart of snowfall in Korea (The average snowfall from 1971 to 2000; unit: cm) Topography affects snowfall, and it snows a lot especially in areas close to the sea and by a mountain.

Seonginbong Peak.

Back on the mainland, the east region of the Taebaek Mountains also receives a great deal of snow. The annual average snowfall for Gangneung and Sokcho is around 80 cm, higher than anywhere except Ulleungdo Island and the highlands region. It snows a lot on the eastern side of the Taebaek Mountains when the northeasterly wind is blowing, in contrast to Ulleungdo Island, where it is the northwesterly wind that brings the most snow. In late February and early March, when snow has stopped falling in other parts of Korea and the approach of spring is palpable, snow is falling on this east-coast area. The news reports villages isolated by heavy snowstorms, and traffic is controlled to make sure that all vehicles ascending the Taebaek Mountains are equipped for snowy conditions.

The reason for this weather is that the distance between the coastline

Snow on Albong and its surroundings on Ulleungdo Island It is particularly known to snow a lot around Albong peak, north of Seonginbong Peak on Ulleungdo Island. (Ulleung, Gyeongsangbuk-do. Jan., 2000)

and the main ridge of the Taebaek Mountains rarely exceeds 10 km. When the northeasterly wind stacks thick clouds against the eastern side of the Taebaek Mountains, this narrow region receives heavy snowfall.

On the other hand, when the cold northwesterly winter monsoon blows powerfully, a great deal of snow falls on the western coastlines of Jeolla-do and Chungcheong-do. In particular, it snows a lot in the Jeolla region. Quite frequently other parts of Korea enjoy clear, sunny weather while snow is falling in Jeolla.

When I was a university student, I was not well aware of the nature of the snow in the Jeolla region, an ignorance that caused me trouble on several occasions. When I was returning home to Jejudo Island after a long absence, I would take an early express bus from Seoul to Mokpo, anxious not to miss the ferry. The bus would leave Seoul on a perfectly clear day,

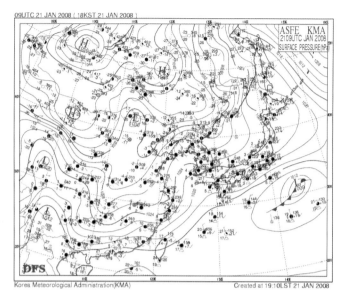

09UTC 21 JAN 2008 (18KST 21 JAN 2008)

ASFE KMA
21 09UTC JAN 2008
SURFACE PRESSURE(hPa)

Korea Meteorological Administration(KMA)   Created at 19:10LST 21 JAN 2008

The surface pressure chart on a day that it snowed in the east coast of Korea There is a chance of heavy snowfall in the east-coast area of Korea when a high pressure air mass stands in the north or northeast of the peninsula and the direction of the isobaric line is developed from the northeast to the southwest. (Korea Meteorological Administration. Jan. 21, 2008)

but near Cheonan, I'd spot one or two small clouds in the west sky. When the bus passed Daejeon, and had just entered the Honam Expressway, there would be more clouds. From there, the number of clouds would rapidly increase the further south the bus headed, and the minute the bus entered Jeollabuk-do, it would start to snow. By the time the bus reached the Honam Tunnel, which passes through the Noryeong Mountains, the expressway would be covered with snow, delaying the traffic. Though I had left Seoul with what seemed like ample time for the trip, by the time I arrived at Mokpo Port, the passenger ferry for Jeju Port would have already left.

As if to make up for those dreadful moments on the Honam Expressway, for some three years, whenever I heard news of snow in the Jeolla region, I

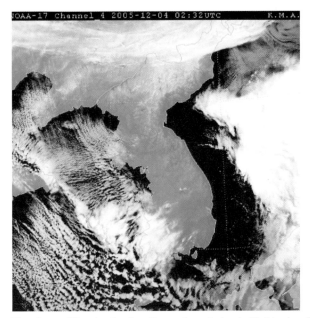

The shape of the clouds when the northwesterly wind is blowing These clouds leave heavy snowfall in the west-coast area of the peninsula. (Source: Image of NOAA (National Oceanic and Atmospheric Administration) / Korea Meteorological Administration. Dec. 4, 2005)

drove along the west coast to Gunsan, and made several field trips between Gunsan and Jeonju. My driving along a road where snow was piling up by the hour worried my family and friends, but I learned many things from those trips. I learned, for example, that the snow that falls when the northwesterly wind is blowing starts from Chungcheongnam-do, while passing Asanman bay. At such times, it doesn't snow in Gyeonggi-do. If one looks towards Gongse-ri from Asanman in Pyeongtaek, one can see that it is snowing in large flakes, although no snow is falling in Pyeongtaek. I also discovered that the snow that falls in Gunsan is different from that in Jeongeup or Jeonju, even though all three cities are in Jeollabuk-do. The snow of coastal area is due to the sea-effect, while the snow that falls on the highlands and the surrounding mountain areas is due to forced ascending

Ginseng farms after a snowstorm Advance preparations can most of the time minimize damages caused by snow as seen in the difference between the two above ginseng farms. (Buan, Jeollabuk-do. Dec., 2005)

by totpgraphy. Making these discoveries and coming to understand these phenomena, I felt compensated for the ordeal I had been through nearly twenty years earlier.

The snow in the Jeolla-do also differs from that in the east coast. In the former, it snows often when the northwesterly winter monsoon is strong. Therefore, the residents are prepared for frequent snowfall. But when it comes to heavy snowstorms, they are relatively negligent. When a massive snowstorm hit the Jeolla-do in December 2005, it shook the whole region. It caused massive damage, with collapsed homes, and destroyed greenhouses. Unlike rain, snow accumulates as it falls, causing damage under its weight. But if preparations are made, such damage can be minimized. In the 2005 snowstorm, the facilities of most ginseng farms collapsed, but there were some neighboring ginseng farms that remained untouched.

On Jejudo, when the northwesterly wind is blowing strongly, it snows along the northern slopes of Hallasan. Those villages between 100 and 200 meters above sea level are called mid-mountain villages. Those villages are subject to a lot of snow, and villagers have their own word, *nunbung'ae*, to describe a snowstorm with raging winds. Encountering one of these local blizzards in a vast field, it is easy to become completely disoriented. Growing up in that area, I knew people who had lost family members to *nunbung'ae*. After my childhood experiences of those fierce blizzards, the silent snowfall in Seoul came as a real surprise.

12

Different winds
for different seasons

The direction of the wind changes in Korea according to the season. Such season-specific winds are called monsoon. In the winter, there are many days when the northwesterly winter monsoon blows powerfully because the air pressure of the Siberian plain is high. In the summer, the continent heats up quickly, but the sea warms up only gradually, so the air pressure of the Pacific is high, and the southeasterly or the southwesterly summer monsoon blows from the Pacific.

n midsummer, there were many times when I would wait eagerly for the wind. When I was in elementary school, during the time just before summer vacation, I couldn't feel a single breeze as I walked home from school. I tried whistling, hoping it would bring wind, but it was no use. Yet my discomfort was nothing compared to what my parents and all the grownups in the village must have felt. Farmers who sit weeding in a muggy bean patch, with baking hot ground below and a scorching sun overhead, start yearning for even the slightest breeze. At such times, it was the grownups of the village who whistled. When a gentle breeze happened to blow, you could completely forget the hot sun and muggy ground heat. It was a most welcome breeze, like a sip of water after some sweaty exertion. And each such breeze encouraged the villagers to believe that their whistling brought the wind.

But whistling in midwinter earned a quick scolding from the older folk

A farmer sits weeding in a muggy bean patch with baking hot ground below and a scorching sun overhead, yearning for even the slightest breeze. (Jeju, Jeju-do. Aug., 2007)

in the village. Jeju islanders have no fondness for the winter wind; it blows constantly, with no need for whistling. Indeed, the cold winter wind is sometimes unbearable on Jejudo Island. That icy wind comes as an unpleasant surprise to tourists who visit Jeju in winter, expecting the pleasures of a warm southern island. It's true that the temperature rarely drops below freezing point on Jeju, even in the middle of winter. Yet most people visiting the island for the first time in winter cry out the minute they arrive, "How can it be this cold on Jeju?" Those tourists who arrive unprepared for the weather usually return home after their rude awakening. Natives of Jeju think of Seoul as a frightfully cold place, yet Seoulites find the cold of Jeju unbearable. The culprit is that unceasing northwesterly wind.

The fierce wind of winter

The direction of the wind changes greatly in Korea according to the season. Seasons are characterized by what is known as monsoon. As Korea is located between the continent of Eurasia and the Pacific Ocean, the direction of the wind changes according to the difference in temperature between those two bodies. In the winter, there are many days when the cold northwesterly winter monsoon blows powerfully because the air pressure of the Siberian plain is high. In the summer, the Pacific Ocean has the higher air pressure, so the southeasterly or southwesterly summer monsoon blows from there.

The reason for the powerful wind in winter is the great difference in temperature between the continent and the ocean. On the Siberian plain, the temperature goes down to nearly minus 50°C, while in the Pacific Ocean, it is over 20°C. Moreover, the sea near Korea rarely goes below minus 10°C. This marked difference in temperatures between the Siberian plain and the Pacific is what generates the strong winds.

Winds start from differences in temperature. At cold places, the air

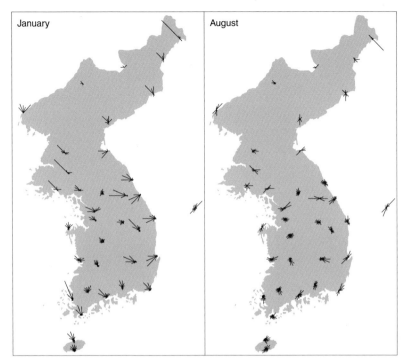

A wind rose diagram for January and August in Korea The diagrams indicate the direction of the winds blowing on the peninsula, which are mainly the northwesterly in the winter and the southeasterly or the southwesterly in the summer.

accumulates, and as the density rises, an anticyclonic system develops, with high air pressure. In warm places, on the other hand, the heated air becomes light and ascends, so the air density is low, causing a cyclonic, low pressure system. The air moves from the dense, high pressure, anticyclonic system towards the low pressure system. And that movement of air is what we experience as wind. The wind is strongest in midwinter, the period when the Siberian plain is at its coldest, which is also the coldest time of year on the Korean peninsula. Thus Koreans know to expect what they call the Great Cold around January 21.

Standing on the sand dunes at Hyeopjae Beach, in northwest Jejudo Island, on a cold, windy winter day, you immediately comprehend the meaning of wind. Even standing still, you feel something constantly hitting your face. It is the fine sand from the beach, whipped up by the wind. The dune is covered with a sheet of vinyl to prevent the sand particles from swirling into the air, but to no avail. At least, there is a small island, Biyangdo, in front of Hyeopjae Beach, to provide a bit of protection from the northwesterly. But Gimnyeong Beach, in the northeast, has no such shielding island; standing there, you feel the full, relentless power of the wind. But growing up in the mid-mountain region, I was unaware of force of Jeju's winds until I was older.

When it comes to winds on Jeju, there is a big difference between the mid-mountain area and the coast. On the same day that the wind on the

Hyeopjae Beach in winter The sand on Hyeopjae Beach is covered with a sheet of vinyl all winter to prevent the sand particles from swirling into the air, and from being carried towards inland by the fierce northwesterly winds. The island in the distance is Biyangdo. (Hyeopjae Beach, Jeju-do. Jan., 2006)

beach has been flinging sand in your face, if you go up to the mid-mountains, you will find it mild. That's because the wind is calmer and milder. Indeed, people who live by the coast say they can hardly feel any wind at all when they visit the mid-mountains.

Before Jeju's beaches open for the summer tourist season, you can observe an interesting sight that testifies to the force of the wind. The managers of most of the beaches on the northern coast make an effort to prevent the sand from blowing away in the winter. Nonetheless, a huge amount of sand is blown inland. Therefore, in June, work needs to be done to make up for the sand that has been stripped from the beaches. To that end, large dump-trucks are mobilized to carry sand back to the beaches. The winds carry sand as far as 2 or 3 km inland, where it piles up on farmers' fields. Naturally the owners of those fields have to remove the sand

In June, work needs to be done to make up for the sand that has been stripped from the beaches. To that end, large dump-trucks are mobilized to carry sand back to the beaches. (Hyeopjae Beach, Jeju-do. June, 2007)

The winds carry sand as far as 2 or 3 km inland, where it piles up on farmers' fields. Naturally the owners of those fields have to remove the sand in order to grow their crops. (Jeju, Jeju-do. Jan., 2006)

in order to grow their crops.

Looking westward during a blizzard on the west coast, you can witness the full force of the northwesterly wind. Even a person who has grown up in an area known for its cold weather will probably question their hometown's claim to fame once they've experienced the cold wind in the open plains of Gimje in midwinter. And it's colder still along the coasts that are even slightly further north.

On the west coast, even the sea pine trees, famous for their strength in the face of sea winds, cannot fully resist: their branches are all bent away from the sea by the force of the wind. Trees with broader leaves are even more prone to the shaping effects of the wind. The trees seen at the shrines of seaside villages are mostly either hackberry or zelkova. These trees are less resistant to salt than sea pines are, so they grow up twisted and

On the west coast, even the sea pine trees, famous for their strength in the face of sea-winds, cannot fully resist: their branches are all bent away from the sea by the force of the wind. (Mongsanpo Beach, Chungcheongnam-do. Dec., 2007)

deformed by the strong, salty winds. There are many such wind-sculpted trees to be seen on the northern coast of Jeju and on islands off Korea's west coast.

Spring is the driest season of the year in Korea. When the air is dry, even a slight difference in heat between regions can give rise to a strong wind. Therefore, even far inland on a large continent, when conditions are dry, fierce winds can blow. So, there are gust in the inland areas of Korea during spring. These can lead to the scary experience of driving down a highway on a lazy spring afternoon and suddenly feeling your car buffeted by a side-wind. However, the wind rapidly weakens as the sun goes down. That is because there is less difference in temperature among regions after sunset. So the monthly average wind speed in April in the inland region is not all that high, but strong winds are frequent during the dry part of the day.

Typhoons: the heralds of autumn

Each year, between late summer and early autumn, one or two typhoons pass through Korea or the areas nearby. They rarely pass through quietly; often they leave considerable damage behind.

Caught in the grip of a prolonged drought, some might be eager for the typhoons. But for most people, typhoons are objects of dread. And they are especially dreadful to anyone trying to make a living at a summer beach resort. A beach that is crowded with vacationers will be instantly emptied by the forecast of a typhoon. After a typhoon has passed through, the sky may be gorgeously blue, and the sea may take on the color of jade, but spread out on the white sand will be far more litter than tourists. Such is the power of typhoons to push visitors away from the seaside.

The sudden absence of vacationers from the white sand beaches around August 15 marks the arrival of a typhoon, indicating the change of season: autumn has nearly arrived. Of course the afternoons are still scorching hot,

A beach that is crowded with vacationers will be instantly emptied by the forecast of a typhoon. (Muchangpo Beach, Chungcheongnam-do. Aug., 2007)

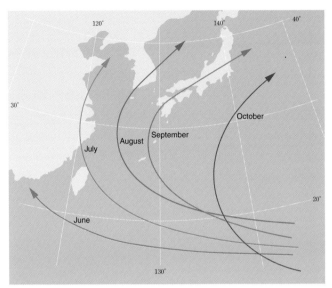

Typhoons can affect Korea from early summer through autumn; however, they directly approach the peninsula in mid to late August.

even after a typhoon has just passed, but most people can sense a change in the atmosphere: after sunset, the air is getting colder with each passing day.

Then why does the season change after a typhoon has passed by? Typhoons are a kind of low pressure system. The earth's air pressure zones are formed in the air over the cold, polar regions and over the hot tropical regions. Near the poles, the air year-round is compressed close to the earth's surface, developing a high pressure, anticyclonic belt. In the area near the equator, on the other hand, the extreme heat causes the air to be less dense, to ascend, and to develop a equatorial low pressure belt. Many small vortexes form in the equatorial low pressure belt. Most dissipate, but the ones that last longer gradually develop into typhoons. These typhoons move from the tropics into the mid-latitude regions, carrying with them the tropical heat and moisture.

When such a typhoon approaches Korea, it passes between the North

Pacific anticyclonic system centered in the south, and the Siberian anticyclonic in the north. That is, typhoons move along the edge of the North Pacific anticyclonic system, following a parabolic course. A typhoon's course is shaped according to a dynamic determined by the relative strengths of the two anticyclonic systems. And during August, that dynamic steers the typhoons directly towards Korea. When a typhoon moves north, it pushes the force of the North Pacific anticyclonic system far away from the Korean peninsula. Therefore, after a typhoon has passed by, and the temperature gradually becomes cooler, because the power of the

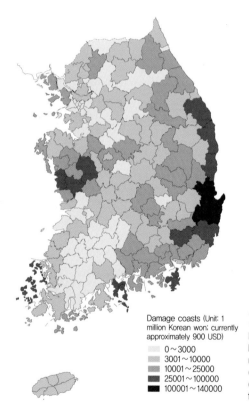

Damage coasts (Unit: 1 million Korean won; currently approximately 900 USD)

- 0~3000
- 3001~10000
- 10001~25000
- 25001~100000
- 100001~140000

Damages from typhoons throughout Korea Damage costs are generally higher in the east coast (from record-breaking torrential rains in the mountains) and in the southeast (industrial) area of the peninsula. (Source: Annual report on disaster, from 1991 to 2000)

North Pacific anticyclonic system drive out from Korea peninsula. In that sense, typhoons can be called the messengers of autumn.

Typhoons travel fairly fast. If the evening television news shows footage of the residents of Korea's islands and coasts fearfully eying signs of an approaching typhoon, the following morning's broadcast will report that the typhoon has already passed, its business on the peninsula concluded.

Typhoons near Korea travel at about 40 km per hour, so in the course of a single night, a typhoon that was south of Jejudo Island could easily make its way far out into the East Sea. The day after a typhoon has passed, the sky is a clear blue and visibility is great, very pleasant and refreshing for residents of Korea's inland areas. People living on the islands and coastal areas, however, have no leisure for admiring the sky; they're too busy repairing the damage that the typhoon has left behind. The region of Korea most subject to damage from typhoons is the southeast coast. The area of greatest damage varies according to the path of each typhoon, but in general more typhoon-related damage occurs on the east coast than the west. Based on the direction a typhoon travels, the area to the right is called the dangerous semicircle, while the area to the left is known as the navigable semicircle. Inside a typhoon the wind blows counter-clockwise, and the prevailing wind around Korea is westerly. Therefore, on the right-hand side of the direction the typhoon is moving, the southwesterly force of the typhoon's own wind is enhanced by the force of the prevailing westerly wind. On the other hand, on the left-hand side, the typhoon generates a northeasterly wind, which loses strength when it encounters the prevailing westerlies.

Given the usual path of typhoons approaching Korea, the east-coast region is generally within the dangerous right-side semicircle. When a typhoon passes far to the east of the peninsula, it has very little effect on Korea, but if it passes over the Yellow Sea, or directly through the peninsula, then the east coast falls within the dangerous semicircle.

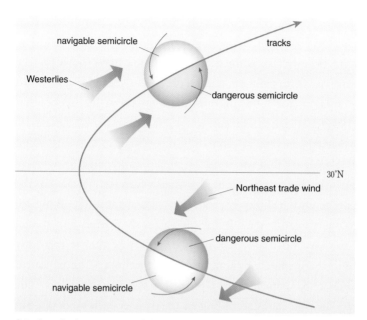

A typhoon's dangerous semicircle and navigable semicircles On the right-hand side of the direction the typhoon is moving, the southwesterly force of the typhoon's own wind is enhanced by the force of the prevailing westerly wind. On the other hand, on the left-hand side, the typhoon generates a northeasterly wind, which loses strength when it encounters the prevailing westerlies.

Moreover, when Korea is under the influence of a typhoon, a northeasterly wind is blowing in most parts of the country. When that northeasterly meets the Taebaek Mountains, and the air is forced to ascend, large amounts of rainfall. On top of the strong winds brought by the typhoon, that heavy downpour of rain can double the damage. In contrast, unless a typhoon passes over the Yellow Sea, the west coast falls within the relatively safe, left-side semicircle. Furthermore locally, heavy rainfall is unlikely on the west coast, given the prevailing northeasterly wind.

Typhoons cause a lot of damage on the mainland with their strong winds and heavy rainfall. But they can pose even greater dangers at sea.

A port is filled with fishing boats by the forecast of a typhoon. (Taean, Chungcheongnam-do. Aug., 2007)

Typhoons bring high waves and strong winds that imperil any ships in the area. In addition, a passing typhoon causes a rapid drop in air pressure, which leads to a sudden rise in the sea level. This puts nearby coastal areas at risk from tidal waves. If a typhoon should hit the mainland at high tide, the coastal area can suffer massive flood damage. This was the tragic scenario on September 17, 1959, when Typhoon Sarah struck.

The vortexes produced in the tropical seas grow, and if the maximum wind speed at the center reaches 17 mps, it is called a typhoon and given its own name. The names for typhoons are submitted by fourteen Asian-Pacific countries that are subject to their effects, and the ten names provided by each country are used in sequential order. South Korea submitted ten names, including Gaemi (ant), Nari (lily), Jangmi (rose), and Noru (roe deer). North Korea also submitted ten names, including

A satellite imagery of Typhoon Rusa approaching from the Pacific Ocean southeast of the Korean peninsula. (The dark dot in the center of the spiral of clouds is the typhoon eye.) (Korea Meteorological Administration. 21:00 hrs, Aug. 27, 2002)

Kirogi (wild goose), Toraji (Chinese balloon flower), Kalmaegi (seagull), and Meari (echo). So, a total of twenty Korean words are being used for typhoon names. However, giving Korean names to a typhoon may not be an effective way of gaining international goodwill, since typhoons that cause severe damage earn the resentment of the people affected by the natural disaster. The typhoon names Maemi (cicada), submitted by North Korea, and Rusa (deer), by Malaysia, are good examples. These two were among the most destructive typhoons in recent years, claiming many lives. Koreans experienced little national pride as victims sighed in the aftermath of super-typhoon Maemi.

The agonies of a soft, gentle breeze
One winter, I took some students from Seoul to Jejudo Island for a

temperature-observation field trip. We arrived at an orchard where, gazing up at the night sky, we were amazed by the panoply of stars overhead. Even the local villagers were impressed by the celestial display. Perhaps the environmental light-show that had so excited the students is to blame for the hardships of the following day. Blame can also be assigned to the students' unfounded belief in 'warm Jeju'.

In any case, by the next morning, all the students had caught colds. The real culprit was the soft, gentle breeze, which was actually the cold air that slowly descends from Hallasan. The thermometer in the orchard recorded a minimum temperature of minus 10°C for that night, and all the students had fallen asleep in that cold air. Jeju natives, long aware of that breeze and its chilling effects, have given it a special name: *nareut*, meaning a soft, gentle breeze, that is nonetheless powerful.

When Korea is in the grip of a severe cold spell, a familiar weather forecast formula goes, "All regions will experience sub-freezing temperatures, except for Jejudo Island." That oft-repeated formula causes most Koreans to believe that conditions are warm and sunny on Jeju. However, that 'above freezing' description really only applies to the windy Jeju City and the area south of Hallasan. In all the places that are exposed to the *nareut* that descends from the mountain, temperatures often drop below freezing.

During a cold spell in January 1990, many of Jeju's tangerine orange trees froze to death because of the *nareut* from Hallasan. Although dense windbreak forests had been planted to block out the cold northwesterly, these wind-blocks actually contributed to the death of the trees. When the cold *nareut* came down from the mountain, the windbreak forest acted as a dam, preventing the cold air from passing further. Thus, the *nareut* that blew all night from the mountains was backed up as a cold air lake. Most of the orange trees in the area 'protected' by the windbreak died. To understand what a severe shock this was to the islanders, it is important to

More orchards use nets, which weaken the northwesterly wind, but let the cold *nareut* breeze from the mountains blow through, without blocking in the cold air. (Seogwipo, Jeju-do. Nov., 2007)

realize that those tangerine orange trees had been seen as their salvation: a new and profitable agricultural product that had lifted many farming families from poverty.

Some residents said that their sadness on losing the orchards was equivalent to the experience of losing family members during the April 3 Jeju Resistance. The reference is to a historical event that left deep emotional scars: a civil conflict and military repression between 1948 and 1954, during which nearly all Jeju natives lost at least one family member. At any rate, taking the destruction from the *nareut* as a lesson, the windbreak facilities today have completely changed. Rather than planting trees close together to make a dense forest, the planting pattern is more scattered, and more orchards use nets, which weaken the northwesterly wind, but let the cold *nareut* blow through, without blocking in the cold

The location of green tea plantations on Jejudo Island are exposed to the cold breeze blowing from Hallasan, so there are stainless steel pinwheels installed, neatly spaced apart. (Seogwipo, Jeju-do. May, 2006)

air.

The green tea plantations mid-way up the slopes of Hallasan have a feature that isn't found on Korea's mainland green tea plantations in Boseong, Jeollanam-do. Around the farm, there are stainless steel pinwheels installed, neatly spaced apart. These devices are mounted throughout the plantation to prevent the cold air of the *nareut* breeze from gathering around the tea plants. If the cold air from the *nareut* remains on the plantation, the tea leaves suffer from the freezing conditions and become difficult to pick, greatly reducing their market value. The green tea plants in Boseong grow on the side of a hill, so there is no need for such pinwheels. The cold air from the high mountains descends and then disperses into the area further down the hillside. In any case, both Jeju and Boseong green tea plantations are very popular destinations for Korean

When villages are located in valleys, the houses are not built at the bottom, but rather somewhat further up the hill. (Hapcheon, Gyeongsangnam-do. Jan., 2003)

tourists.

In the deep valleys on a windless night, you are certain to feel a cold wind gradually flowing from uphill, like the *nareut* breeze on Jejudo Island. You would be ill-advised to build a house at the bottom of such a deep valley. There can be nothing healthful about sleeping in a thick mass of cold air every night. Hence when villages are located in valleys, the houses are not built at the bottom, but rather somewhat further up the hill.

On trips to farming villages, I am often told of the low productivity of orchards located at the bottom of a valley. There are some farmers who try to convince themselves that the fault lies with the trees. But hearing me explain about the massing of cold air, most of them soon accept that the problem lies with the location: clearly the bottom of a valley is a poor place to plant an orchard. But no farmer wants to let good land go to waste, and

At Yesan, Chungcheongnam-do, the locals have made very good use of climate. At the bottom of the valley, where cold air collects at night, they have planted rice, while the apple orchards have been planted on the hillsides. (Yesan, Chungcheongnam-do. Oct., 2006)

such areas are suitable for growing crops during the summer.

Yesan, in Chungcheongnam-do, is famous for the Oga apples grown there. Looking carefully at the surroundings, you can see that the locals have made very good use of nature. At the bottom of the valley, where cold air gathers at night, they have planted rice, while the apple orchards have been planted on the hillsides. This strategy minimizes damage from the cold air. This arrangement is clearly the product of practical knowledge gained from long years of experience. Farmers who ignore the patterns of nature in an effort to make use of every available bit of land will definitely end up paying the price.

The soft, gentle northeast wind dries up the land

Most Korean geography teachers tell their students that the residents of the area west of the Taebaek Mountains use the word *nopse* to designate the wind blowing from the northeast. And when Koreans think of this area, the town of Hongcheon comes to mind. High school textbooks mention *nopse* when comparing the temperatures and relative humidity of Hongcheon, to the west, and Gangneung, to the east, of the Taebaek Mountains. I once went on a research trip to Hongcheon for a few days to see how the local residents regarded the *nopse*.

In the course of my research, I made a startling discovery. Among the elderly natives of the region I met, not a single one knew what the *nopse* wind was. I made it a point to visit the local old-age facility, but no one there knew anything about the *nopse*. On the other hand, the young people I spoke with knew all about it. However, they informed me that they only knew about *nopse* because they had learned it in geography class! The claim that locals call the hot and dry northeasterly wind *nopse* simply isn't true. I didn't know whether to laugh or cry. Still, whatever its etymology, today *nopse* is taught as the word for the northeasterly that has become warm and dry by descending the lee side of a mountain.

The damp northeasterly wind ascends when it meets the Taebaek Mountains. As it rises in altitude, the temperature falls, and when it reaches the dew point, condensation occurs. At this point the heat released when water vapor condenses works against the general trend of air to be cooler at higher levels. So, whereas air that has not yet condensed drops 1°C in temperature for every 100 meters of increased elevation, once it has condensed, it is only 0.5°C colder for every 100 meters of altitude. Once the air has climbed over the Taebaek Mountains, the clouds evaporate as it starts to descend the slope, and the temperature rises 1°C for every 100 meters that it descends. As a result, the air to the west of the Taebaek Mountains is higher in temperature and lower in humidity than the air to the east of the mountains, before it has crossed the range. The higher the temperature, the more water vapor can be contained in air, so even if the

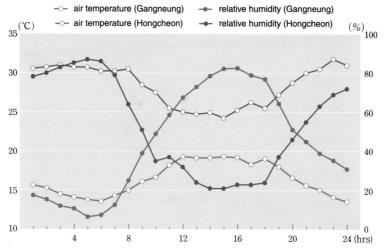

air temperature (Gangneung)   relative humidity (Gangneung)
air temperature (Hongcheon)   relative humidity (Hongcheon)

(°C)

(%)

35

100

30

80

25

60

20

40

15

20

10

0

4   8   12   16   20   24 (hrs)

The changes in temperature and relative humidity in the east and the west (the windward and the leeward side) of the Taebaek Mountains on a day the *nopse* blows, the hot and dry northeasterly wind.

amount of vapor remains constant, when the air temperature rises, the relative humidity falls. This process of air warming and drying as it descends the lee slope of a mountain range is known as the foehn effect.

The *nopse* is the result of the foehn effect, and is regarded in Korea as a rather undesirable phenomenon. The *nopse* is most likely to appear between May and the end of the rainy season. During that period, if you notice a clear sky and a slightly cool breeze when you leave home in the morning, most likely there will be a *nopse*. People from Seoul can tell such a day immediately in the spring by the incredibly clear blue sky, something otherwise never seen in that smoggy city. But it is only nowadays that people can feel excited and refreshed at the news of a *nopse*.

What would it have been like when the great majority of Koreans made their living through agriculture. A *nopse* effect lasting several days would have been most unwelcome. If the dry wind continued to blow, the

A drought poses more threat to farmers growing vegetables than rice. Sprinklers are run to water the fields when a dray spell persists. (Seogwipo, Jeju-do. June, 2006)

farmland would start to dry out. What's worse, the wind arrives right after the dry weeks of spring, so it would have made it difficult to transplant rice. Thus, the *nopse* was traditionally regarded as the farmers' enemy.

In Hwanghae-do, it is said that there was a traditional farming technique in the dry season that involved tramping on the fields. Compressing the soil was a means of exploiting capillary action in order to draw buried moisture up towards the surface. Many years ago, when I was a young child on Jejudo Island, each spring, after sowing a crop of fox tail millet, farmers would release a large herd of horses onto the field, and make the animals tramp the soil. So, in order to be regarded as wealthy, a farmer had to own at least a dozen horses. Farmers who had no horses used a tool called *namtae* instead. It was a heavy, round log studded with rows of wooden blocks. Dragged through the filed by a single horse, the *namtae* achieved the same result as dozens of horses stomping with their hooves.

Today, the only place you can see a *namtae* is in a museum.

Nowadays a *nopse*-induced dry spell no longer poses a threat to agriculture, whether east or west of the Taebaek Mountains. That's thanks to the invention and installation of sprinklers. Farms throughout Korea are now well equipped with irrigation facilities. That is why, although you often hears news of flood damage in Korea, you almost never hear news reports of droughts caused by the hot and dry northesterly, *nopse*.

To know an island, get stranded there for ten days

I once made a visit to Ulleungdo Island to see its famous snow. On my fourth day there, the ferry services were cut off; they didn't resume until ten days later. The ferries that run between the mainland and Ulleungdo Island are high speed craft that don't operate in rough seas. So when there is a storm warning, all sailings are cancelled. Worst of all, however, is that the two mainland ports that connect to Ulleungdo Island – Pohang (in Gyeongsangbuk-do) and Mukho (in Donghae, Gangwon-do) – are covered by two different area from the Korea Meteorological Administration. Consequently, the ferries out of Ulleung can only resume running when the storm warnings have been lifted for both ports.

On the first day that the ferry service was cut off, most people seemed to be happy about the unexpected opportunity to extend their stay on the island. Besides, there was no basis for complaint, since storms are acts of nature. And facilities for rest and relaxation had been provided. So an attitude of patient acceptance prevailed, not just in my group, but among the other tourists as well. However, moods changed as first one day, then another went by. Patience gave way to complaint and recrimination. "Why didn't we take the earlier boat?" was an oft-heard reproach. That lasted for about two days, after which the whole village quieted down. And a week later, the village was totally silent. Once every two hours, I walked out to Dodong pier; there was nothing in sight. And so passed another few days.

Jeodong pier upon a storm warning
(Ulleung, Gyeongsangbuk-do. May, 2007)

"So this is what life is like on an island," I told myself. Things were just as quiet on the neighboring Jeodong pier, where the dock was filled with fishing boats that couldn't go out to sea. The fishermen themselves were all either mending their nets or just resting near the port.

The tenth day brought word that the ferry was on its way. The news spread unbelievably fast. Out of curiosity, I visited the pier again. The place was already crowded with people, although there was no sign of an incoming boat. Taxis were waiting in line, and inn owners were standing with welcoming signs in their hands. It wasn't until more than two hours after the expected arrival time that the sound of the boat whistle was finally heard. Ever since then, I've been able to understand and empathize with the joyous, relieved expressions on the faces of people stepping off a ferry onto the mainland after a visit to an island.

Jeju Port With an increase in the transportation of passengers and cargos between Jejudo Island and the mainland, port facilities are being expanded further out towards the sea. (Jeju, Jeju-do. Jan., 2008)

Jejudo Island was different, however. On visits there, I never thought of it as an island. Even when I arrived by ferry, the idea never occurred to me. I was simply going to a place called Jeju. But that changed during one visit. It was a spring holiday, and a storm warning was issued on the day I arrived by ferry. I knew I would have to return to Seoul soon, but the gigantic ferry could not leave the port. Watching the threatening waves crashing against the huge breakwaters, I was forced to the realization that Jeju is indeed an island. There was no denying it. The storm-tossed sea was a completely different beast from the calm Jeju waters I was accustomed to.

When a wind and wave warning is issued on an island, everything becomes silent. Outside the seawalls, big waves roar as though they would devour everything. But inside the breakwater, in the sheltered harbor, all is tranquil and quiet. The people near the ferry terminal have their eyes fixed on the water in the harbor. Seeing nothing but gentle swells, it's easy for

The small Chuja Port upon a storm warning (Chujado, Jeju-do. April, 2008)

tourists to think that the sea has calmed down. Naturally people who live near a port develop a more finely tuned sensitivity to wave conditions.

Several days after the ferry service had been shut down, while I was conducting a field survey near Chuja Port, a voice came through an outdoor speaker announcing that the wind and wave warning had been lifted. The sound of those words was sweeter to me than a shower of rain after a long drought. Such is life on an island.

13

Fog and frost are local phenomena

When there is large difference between daytime and nighttime temperatures, fog is likely. The air close to the ground starts to cool down at night, but, due to temperature inversion, the air further up gets warmer. The inversion is the main cause of fog. Among weather phenomena, frost has the most significant direct impact on agriculture. The agricultural growing season is defined in terms of frost-free days: the period between the last frost day and the first.

One winter, I sat on the bank of the Namdaecheon stream, in Gangneung, admiring the dense fog. I had no negative thoughts at all about the fog; quite the contrary, it struck me as romantic. But a few years later I did my military service as a meteorological officer in the Air Force. From then on, fog presented itself to me in a totally different way from that dreamy vision by a stream in Gangneung. When fog would start to appear at dawn, it felt like a creepy, hateful person approaching me. When I was on duty at the base, I'd start my day by opening the window to look at the sky. Naturally, it was to check for fog.

From then on, I never regarded fog with the aesthetic admiration I felt at the Namdaecheon stream. More and more, I began to view fog from a realistic perspective. When I became a professor, the first M.A. thesis for which I served as advisor was about fog. Fog became a practical subject and

Fog in Gangneung In the east coast of Korea, fog settles in along with the rain. The fog that appears along Korea's east and west coasts is mostly sea fog. (Gangneung, Gangwon-do. July, 2008)

part of my academic life: an aspect of the climate to research and to explain to students. As an unfortunate result, I no longer experience fog in terms of its beauty.

From my high school Korean Literature textbook, I remember the story about a man who, looking up at the moon on a cold, quiet, frosty night, was stirred by memories of the hometown he had left long before. Perhaps, as in that story, frost may be regarded as more emotionally evocative than fog. Nonetheless, frost is the more concrete, down-to-earth weather phenomenon.

The Buddhist scriptures contain this line: "Would I dare violate a single rule from the Buddhist precepts that are as strict as the frost?" That suggests that frost is associated with great severity. Certainly anyone engaged in farming is well aware of the dangers of frost. From long ago, farmers have known how to avoid frost while growing their crops.

Frost set on autumn leaves Gentle though it may seem, frost causes fatal damage to any crop.

A hoarfrost-whitened tree stands near a river and a valley. (Yeongyang, Gyeongsangbuk-do. Jan., 2007)

Trees covered with frost are wonderfully picturesque. The white ice crystals that are loosely deposited on tree branches are a kind of hoarfrost. Weather offers nothing more beautiful than the sight of a hoarfrost-whitened tree standing firmly against the cold wind in the early morning. And the white rime on a tree by the riverbank in sunlight seems to emit a misty, mysterious, spiritual light: a memorably scenic image for anyone passing by.

Is fog really something beautiful?

Around the time when the heat of August starts to lose its searing intensity, dense fog sets in at early dawn. The fog starts to rise in low valleys or along the riverside, and gradually spreads throughout the village. Although the place and time of its appearance varies with the locale, fog is the characteristic element of autumn weather in Korea.

For fog to set in, the temperature difference between day and night must be wide. If the temperature drops below the dew point at night, water vapor condenses and it becomes foggy. At night, the ground receives no energy from the sun, but the surface continues to disperse heat into the air, and it cools off slowly. If this process continues all night, the temperature of low altitude is lower than high altitude. This phenomenon, known as temperature inversion, occurs more often on clear and windless nights. Since Korea is often affected by a migratory anticyclone in autumn, there are more clear and calm nights than during other seasons. Therefore, temperature inversion occurs frequently, and there are many foggy days.

Temperature inversions occur because the earth's surface cools quickly. When the surface is covered with snow, or when it is frozen, cooling occurs all the more rapidly. The reflectance of snow is high, so it reflects most of the sunlight it receives during the day back into the sky. Therefore, once the sun has set, the air right above the ground starts to cool very rapidly.

Meanwhile, an unfrozen lake absorbs and retains heat during the day. In

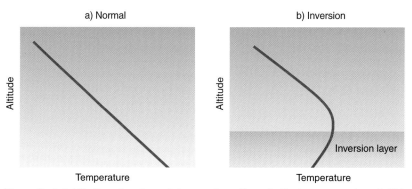

The vertical distribution of surface-air temperature Normally, the temperature drops in high altitudes, but when the air is warmer higher in altitude, the phenomenon is known as temperature inversion.

Surrounded by high mountains that gather cold air, temperatures drop easily, and the surface of the water quickly freezes in the winter at Paldangho lake. (Yangpyeong, Gyeonggi-do. Jan., 2008)

the winter, it is less cold near such a lake. That is because the lake releases much of the retained heat into the air. But once a lake freezes, circumstances greatly change. A frozen lake is like a mirror, reflecting all the solar energy that shines on it during the day. Therefore, in the vicinity of a frozen lake, once the sun goes down, cooling occurs very quickly, just as it does when the ground is covered with snow.

Temperature inversions develop often in basins surrounded by high mountains or in valleys where cooling occurs quickly. At night, cold air from the mountain heights descends to the bottom of the valley. The greater the difference between the top of the mountain and valley, the lower the temperature will be in the valley. Therefore, if an orchard is planted at the bottom of a valley, the trees will be damaged by the stagnant cold air at night.

Most of the well-known old Korean Buddhist temples were built midway up mountains, rather than at the very bottom of a valley. It is not known whether an awareness of temperature inversion figured into the siting of temples, but what is certain is that the temples were never sited deep in valleys. Nothing good can come from living in a place where cold air collects. If you climb a mountain to visit one of Korea's old Buddhist temples when fog has set in the valley, you will find that the temple is located above the fog.

The cooling of the earth surface or the inversion layer caused by the cold air that descends from the nearby mountain hills produce fog. Indeed, the presence of fog is an indicator that an inversion layer has been made. Whether in a city or a valley village, the reason fog settles in is that a temperature inversion is taking place. That is why the air feels slightly cool

Fog set as far up as the temperature-inversion layer On low altitude, fog is set on the ground, but on high elevation, up on mountains or skyscrapers, one is above the fog. The presence of fog is limited to a certain altitude. (Chuncheon, Gangwon-do. Nov., 2007. By Park Chang-yeon)

when there is fog. On a foggy morning, you might be tempted to put on a thick coat upon leaving the house. But you will probably regret the decision later, since usually the weather turns quite warm once the fog has lifted. Generally, fog appears in the inland parts of the Korean peninsula when there is migratory anticyclone on the Yellow Sea. Then, the weather becomes milder as the center of the anticyclone approaches the peninsula. Therefore, on a foggy day, although it feels chilly in the morning, it is better to dress lightly.

This type of temperature inversion layer usually disperses easily. An inversion layer is disrupted when the wind blows or when the ground warms up. In autumn, even the densest fog normally clears by the end of the morning rush hour. That is because by that time, the ground has had time to warm. The sudden gusts of wind that blow at dawn also prevent fog from forming. The airs of the upper and lower layers are mixed as the wind blows. However, when there is thick fog in the deep valleys, and there is a lot of water vapor in the air, the fog does not clear that easily.

One year, I spent an autumn in the lakeside city of Chuncheon, in Gangwon-do. The fog there was beyond belief. The city would be shrouded in a dense fog that didn't lift even in the afternoon. It was nearly suffocating. And on winter nights when the ground was covered with snow, fog would appear again in the early evening.

In addition to temperature inversion, the other important factor in creating fog is water vapor. Thus, most places in Korea with a reputation for fog are in the vicinity of a large lake. When a new artificial lake is made near a meteorological observatory, it generally leads to increased fog. Therefore, whenever plans are announced to build a large dam, there are bound to be protests from local residents. The presence of fog reduces the amount of sunshine, which can have an adverse effect on local agriculture.

Places in Korea with reputations for fog include: Seungju (Jeollanam-do), Yangpyeong (Gyeonggi-do), Jinju, Hongcheon, Andong, Chungju,

Even the thickest fog gradually clears when the sun rises and warms the Earth's surface. (Chuncheon, Gangwon-do. Dec., 2005. By Park Chang-yeon)

Jangsu, Hapcheon, and Chuncheon. All these places are located inland. Finding each of these places on a map reveals the reason for their fogginess. Most are located near a large lake, in an area surrounded by mountains. Especially remarkable are: Seungju, averaging 91.7 foggy days per year; Yangpyeong, with 83.9; and Jinju, with 78.8. Chuncheon averages 60.5 days of fog per year, which would scarcely seem to justify its reputation. However, on days when fog appears in Chuncheon, it lingers much longer than elsewhere, thanks to the areas many lakes, and the high mountains around it.

Fog is much more local than other meteorological phenomena. And on occasion the measurements recorded by a meteorological observatory don't agree with the perception of local residents. In the same city, on the same day, some people will have experienced fog, while others will not have

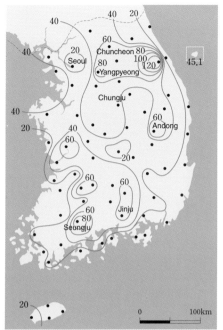

The number of foggy days per year in Korea Places in Korea with reputations for fog are mostly located near a large lake, and in an area surrounded by mountains.

encountered any. Since the development of fog depends on a layer of temperature inversion, its presence is limited to a certain altitude. Consequently, whether any given part of a city experiences fog on any given day depends on its altitude. I remember one autumn when I was commuting to work in Seoul. Every morning I had to cross the Cheonho bridge. From Amsa-dong all the way to the bridge, I would be driving through thick fog. But once I crossed the Hangang river and started the ascent towards the Walker Hill Hotel, on many days there would be no sign of fog at all. In fact, across the hill, it would be all sunshine.

And once I stayed overnight near Yangpyeong to make some temperature observations. The fog was very impressive near where I was staying: a condominium composed of three buildings along a road on a high slope.

Looking from the hillside, the lowest building, furthest down the hill, was completely hidden in the fog; on the second building, midway up the hill, some of the floors were visible above the fog; and the third building, at the top of the hill, was completely above the layer of fog. Paying the same rate for their rooms, some guests would be sleeping in the unhealthful fog, while others would spend the night comfortably above it.

In coastal areas there are fewer foggy days, but once fog sets in, it doesn't clear quickly. That is because coastal fog is of a different type from what occurs inland. The inland fog does not form if there is wind, but the fog along the coast is actually produced by wind. If warm air from the sea meets the cool air that covers the land, the difference in temperature can lead to fog. Known as sea fog, this is a common feature along the shores of England and the east coast of North America.

The timing of sea fog is quite irregular compared to the fog that forms in inland valleys. It depends on the direction of the wind. Generally, sea fog settles in when the wind blows from the sea, and clears away when the wind blows back towards the sea. So, given the right wind direction, sea fog can appear in the middle of the night, as well as in broad daylight. The fog that appears along Korea's east and west coasts is mostly sea fog.

I once worked for a year at a university on Jejudo Island. One night, during my first or second night at my new job, I walked out of the building and discovered that I could hardly make out objects just a few feet in front of me. It was difficult to drive because the beam from my headlights was reflected back by the fog. It seemed as though a ghost was about to pop out. I was suddenly reminded of all the ghost stories about fog and mountain trails that I used to hear, back when I was a child walking 4 km to school. My hair bristled and I felt goose pimples as I tried to direct all my attention to the road. When I finally managed to arrive safely to my home, which were about 30 meters above sea level, there was no fog in sight. The university campus was more than 300 meters above sea level.

Generally, sea fog settles in when the wind blows from the sea, and clears away when the wind blows back towards the sea. (Hongdo, Jeollanam-do. May, 2005)

According to climate data, there is hardly any fog on Jejudo Island; however, this isn't true for the entire island. The meteorological observatory is located near the sea, where, indeed, fog is rare. But at higher altitudes on the island, there is fog nearly every day from spring through the rainy season. On Jeju, spring fog is known to promote the growth of bracken ferns. So on foggy days you can find women picking bracken at the foot of Hallasan. Koreans have long used bracken as an ingredient in many popular dishes, and nowadays there are Koreans who pick bracken as a hobby and regard it as a health food. But not so long ago, bracken was a significant source of supplementary income for farming families on Jeju. Even young children hiked to the mountain to pick the ferns. Bracken like humidity, so foggy places are good areas for picking, and the islanders used

Many people pick bracken at the foot of Hallasan in spring. (Seogwipo, Jeju-do. April, 2008)

to look forward to the spring fog. But today, fog in the mountains has become the main cause of serious traffic accidents on Jeju.

I once had a conversation with an employee of the Jeju Racecourse, who told me that fog is a serious problem at horse tracks. Horses can't run well if they are unable to see ahead. The employee said that the owners of the racetrack were looking for ways to clear fog from the track. My advice was sought. I suggested that they would have better luck putting a roof over the tracks. It is virtually impossible to remove fog from the Jeju racetrack. The fog there is like a cloud. When humid air from the sea reaches the island, a ascending air current develops and forms clouds, which seem like fog at the track. One can only conclude that they chose a bad site for the racetrack. Clouds and fog are both condensed water vapor, and are nearly the same phenomenon. If it touches the ground, it is called fog; detached from the ground, it's called a cloud. The difference in names is largely a matter of

Clouds are formed half-way up Hallasan. Many people hiking on Hallasan may mistake it for fog. And although they are actually clouds, the part of the mountain is called a frequent-fog area. (Hallasan, Jeju-do. Jan., 2008)

perspective, with many equivocal cases. The fog/clouds plague at the Jeju Racecourse is one such case.

Statistically, Korea's foggiest place is the Daegwallyeong Pass, which averages 127.3 foggy days annually: well over one day in three. Fog is so frequent because Daegwallyeong is located on a high mountain. What is observed as low clouds in other areas is regarded as fog at Daegwallyeong – and a very dense one, too. Such fog provides water vapor to plants in the surrounding area, which works better for farms in mountain areas than in the flat lands.

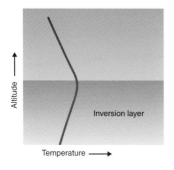
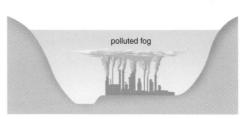

polluted fog

Inversion layer

Altitude

Temperature

Air pollution upon a temperature inversion In layers of temperature inversion, there is almost no air movement and this conduces to the build-up of pollutants from factories and automobile exhaust on the Earth's surface.

People tend towards sentimental feelings whenever they encounter fog. However, fog has many more faults than virtues. Most fog contains pollutants. Fog develops in layers of temperature inversion, where there is almost no air movement. But the same still air that allows fog to accumulate also conduces to the build-up of pollutants from factories and automobile exhaust. Therefore, times when fog is present are usually also times of maximum air pollution. People who believe they are treating their bodies to fresh air by jogging in the early morning fog are greatly mistaken. Exercising in the fog is little different from running past smoke stacks in an industrial area.

If a woman bears a grudge, there will be frost even in summer

There are records from the nineteenth century of frost in July and August in North America and Europe, an after-effect of volcanic eruptions. But in Korea, a frost in May or June is very rare indeed, especially when going by the traditional Korean lunar calendar, where those months generally fall in the middle of summer. Hence the Korean saying, "If a woman bears a grudge, there will be frost even in May and June," conveys at once the terrifying force of a woman's resentment and the intense concern that farming people felt regarding unseasonable frost. Among the phenomena of weather, frost poses the most direct danger to agriculture. When frost attacks a vegetable patch, the result is a mess of withered greens that look like they'd been doused with a bucket of hot water.

Farmers are constantly on guard against frost because it can be fatal to any crop. Tea plants are probably the most frost-sensitive crop grown in Korea, since the marketable product is the delicate leaves. When frost forms on a tea-plant, it kills the cells in the leaves; the damage is fatal, and the crop is lost. The reason that Korea's tea plantations are concentrated in Jejudo Island and in the south coast area of the mainland is that those places have far fewer frost days than elsewhere in the country. The highest quality green tea is *Ujeon*, and the name translates as *Before the Rain*, indicating that the leaves are picked before the first day of spring, which

Frostbitten, withered turnip leaves. (Seogwipo, Jeju-do. Feb., 2008)

Tea trees in the south coast, like in Boseong and in Hadong, are on steep slopes and the cold air flows downhill without stagnating around the plants, which are hardly damaged by frost. (Boseong, Jeollanam-do. May, 2006)

falls around April 20. Apart from Jejudo Island and the south coast, everywhere else in Korea is liable to occasional frost days until the first spring rain. On Jeju, though, there is almost never frost once March has arrived, and by the end of March there are no longer any frost days on the south coast of the Korean mainland.

Frost is apt to appear where cold air remains stagnant for a long time. A place where frost frequently forms is called a frost path. Man-made frost paths can be used to control frost. Since air, like water, flows along a given path, it is possible to direct it artificially. In most of the tea plantations on the south coast, the tea plants are grown on steep slopes. This allows the cold air to flow smoothly downhill without stagnating around the plants. Particularly in Boseong, a county in Jeollanam-do that is famous for its green tea, the plants are grown on such steep slopes that it is difficult for

Tea trees in the south coast, like in Boseong and in Hadong, are on steep slopes and the cold air flows downhill without stagnating around the plants, which are hardly damaged by frost. (Hadong, Gyeongsangnam-do. Feb., 2007)

people to walk among them. And most of the farms in Hadong, Gyeongsangnam-do, known for their wild teas, are at the foot of precipitous mountain inclines.

At a depression, where cold air is likely to stagnate, a small campfire is effective in preventing the temperature from falling below the freezing point. The heat from a fire burning at the lowest point of the depression will create a rising current of air. This in turn causes warm air from the surrounding area to gather near the fire and ascend. As the process is repeated, the air is kept in circulation, preventing the cold air from gathering and lingering in one place, and thus keeping frost from forming.

In the early 1990s, I went on a field trip to area on the east coast of Jejudo Island where tangerine oranges are cultivated. When I suggested to the owner of an orchard that a campfire would help to prevent the accumulation of cold air, he responded, "And then what happens if the fire burns all my orange trees?" I explained that the strategy didn't call for a separate fire beside each tree, but just one carefully placed fire in the orchard. Suddenly tears ran down his face. "If only I'd known sooner, my trees wouldn't have been frozen!" he moaned. I still have vivid memories of that farmer's sorrow. In fact, my encounter with that farmer had a strong influence, years later, on the direction of my research. Nowadays farmers are well informed about damaging weather phenomena and steps that can be taken to counter them, so nobody just stands by hopelessly watching their orange trees freeze to death from the cold air.

There was a time when I thought that an overall rise in temperatures would benefit agricultural production. But upon visiting the Korean pear orchards in Naju, Jeollanam-do, I realized that I was wrong. As average temperatures have risen due to global warming, the pear trees have started to blossom earlier. Some farmers say that the blossoms are now coming more than a week earlier than they used to. The problem is the frost: the date of the last frost has occured more widely. So, it is becoming more common for frost to occur while the pear blossoms are in full bloom. When that happens, the flowers wither before fertilization can take place. This is a new worry for farmers. In any region, it is an agricultural disaster if the last frost day comes late. Young sprouts wither, and farmers can only struggle to limit the damage.

Unlike other weather phenomena, when speaking of frost, the first and the last days of occurrence count more than the total number of affected days. The time between the last and the first frost days is called the frost-free season, and it also determines the growing season for crops. The length of the frost-free season in a given area is important in selecting the type of crops to plant there. If a crop needs more time to grow than the number of frost-free days that area can expect, then there is a heightened risk of frost damage to the crop.

A frostbitten pear blossom (Top left) will not develop into a fruit as the pistil will turn black. (Naju, Jeollanam-do. May, 2007)

Usually, the last frost day announces the beginning of a year's farm season. After mid-March, farming starts on the islands in the south, and gradually moves up north. Of course the garlic and the barley fields, which are unaffected by the frost, are a full, bright sea of green. One can enjoy this early spring scenery in Haenam, Wando, Jindo, Yeongam, Jangheung, and Yeosu in Jeollanam-do, and in Namhae, and Geoje in Gyeongsangnam-do. Haenam is famous for its wintering cabbages. The sight of crops growing in the open fields in winter is unique to Haenam, and sharply distinguishes that county from neighboring areas on the south coast. At most parts of Jejudo Island, farming is possible almost year round. Especially at Seogwipo, in the south of Jeju-do, the winter is so mild that frost almost isn't a factor; even in spring there are almost no frost days.

After mid-April, in most parts of the southern Provinces, farmers

It is possible to grow cabbages even in winter in the warmer south-coast area of Korea. With global warming, it is gradually becoming warmer in areas further north. (Muan, Jeollanam-do. Jan., 2007)

prepare for seeding in the fields, and by late April, planting has started nearly everywhere in South Korea. But it is still too early to plant crops near Daegwallyeong. In that highland area, you don't see farmers in the fields until the mid-to-late May. That stands to reason: the last frost day in Daegwallyeong, on average, is May 13.

The first day of frost, sometime in midautumn, effectively announces the end of that year's farming. Although most crops are harvested before the first frost day, there are some products, like ripe persimmons, that benefit from remaining on the tree for the first frost. The first frost naturally starts from the high mountainous regions. In highland areas such as Daegwallyeong, by October, the fields are all covered in white in the morning. At such high altitudes, vegetable beds are already desolate by late September, and the fields are almost completely ready to greet winter.

A highland farming area in early summer resembles a desert, far from its' famous image of endless green fields. (Maebongsan, Gangwon-do. June 1, 2008)

Although the highland fields look empty, in many places seed potatoes are neatly hidden beneath the soil.

When it is about the time for the first frost along the south coast, the first news reports of snow on some high mountain are coming in, heralding the arrival of winter. Any fruit still left on the trees is called *kkachibap*, literally, bird-feed. Nobody but the farmer can know whether those fruits were deliberately left hanging on the tree to feed birds, or simply because they couldn't be reached by the harvesters. But in either case, they must be a real treat for the birds.

Part 4

Traditional Korean solutions to weather challenges

14

How do people live
in windy areas?

The wind on Jejudo greatly affected the way people lived. This is the most obious in the traditional design of houses. In earlier days, Jejudo houses had low-sloped roofs to minimize wind damages, and they were protected by one of the two types of stone-wall windbreaks – *Ollae* or *Imun'gan* – which block the wind from directly blowing into the house. On the west coast, where the northwesterly winter monsoon blows fiercely, people set up a *kkadaegi*, a thatched fence windbreak, to block the cold.

The wind affects us more directly than any other aspect of the weather. It packs an enormous physical force: if it can't remove your clothes, it can certainly knock you down. When strong winds are predicted, the Korea Meteorological Administration issues a special warning and takes measures to ensure that the people living in the concerned areas make necessary provision. During such alerts, the routines of life are seriously disrupted in places along the coast.

Thanks to being born and raised on an island extremely famous for its strong winds, I was sure I knew everything there was to know about dealing with the wind. However, when faced with a fierce wind in Europe, I found myself powerless. The wind on Jeju is certainly strong, especially on the shore, in winter. When I would hear visitors to Jeju complain that it was too windy to take photographs, I used to reply, "What? You call this little breeze a wind?" But the winds I encountered on the west coast of Europe was another matter altogether. After several days of watching winds that reached speeds of 50 meters per second, I was overcome with fear. In

The special weather statements issued in Korea concerning winds

Unit: m/s (meters per second)

Type of statement	Warning	Alert
Strong winds	Issued upon forecasts of surface wind speed exceeding 14 m/s or upon maximum instantaneous wind velocity over 20 m/s on land. Issued upon forecasts of surface wind speed exceeding 17 m/s or upon maximum instantaneous wind velocity over 25 m/s in mountain areas.	Issued upon forecasts of surface wind speed exceeding 21 m/s or upon maximum instantaneous wind velocity over 26 m/s on land. Issued upon forecasts of surface wind speed exceeding 24 m/s or upon maximum instantaneous wind velocity over 30 m/s in mountain areas.
High seas	Issued upon forecasts of wind speed exceeding 14 m/s for over three hours at sea or when significant wave height is over three meters.	Issued upon forecasts of wind speed exceeding 21 m/s for over three hours at sea or when significant wave height is over five meters.

Source: Korea Meteorological Administration

Wind power is harnessed as an energy resource in the coastal areas, including the northern part of Jejudo Island. (Jeju, Jeju-do. Jan., 2006)

front of an immense natural force like a windstorm, we humans can only feel hopelessly weak.

But in fact mankind has managed to exploit the awesome power of the wind and put it to practical use. In Europe, wind power has long been an important source of energy. In Korea, too, wind power is harnessed as an energy resource in coastal areas, on islands, and in the high mountains. Wind powered generators are not only useful, they are also have an exotic visual appeal. Everyone is attracted to a novel sight, but when will the novelty of wind powered generators wear off? I can't help worrying that some day those huge metal structures will cease being scenic attractions, and present themselves instead as costly headaches.

The winter wind makes the air much colder. When I was in elementary school, in the late 1960s, the penetrating winter wind made the walk to school very difficult. Although the distance was just 3 km, I couldn't walk the whole way without stopping. I would feel so cold that I had to stop at the half-way point to build a small fire in order to heat some pebbles. Then I would finish the walk, feeling the warmth of the stones in my pockets until I arrived at school. All of my classmates had frozen red cheeks after struggling to walk to school through that cold northwesterly winter wind.

The low houses of Jeju

When Koreans refer to Jeju as the "Island of the Three Abundances," they mean that it has a lot of stones, wind, and women. It is no longer the case that women outnumber men on the island, so Jeju can now be regarded as the Island of Stones and Wind.

Hearing where I am from, many people ask, "What is there to see on Jeju?" My standard reply is, "Everything you see on Jeju, from the moment you arrive until you leave, will interest you and reward your attention." Actually, the fascination starts from the moment Jeju comes into sight through the airplane window, and only ends when it slips out of sight on the flight back home. Visitors from abroad will see things on Jeju that exist nowhere else in Korea. It's a marvelous landscape formed by the distinctive rocks, the subtropical temperatues, and the wind.

Jeju's temperature has primarily affected distribution of plants and crops, but the wind has affected the way the inhabitants live. This is most evident in the traditional design of the houses. Looking down through an airplane window at the colorful villages, the whole island looks very peaceful. That feeling of tranquility starts from the low roofs of the modest houses. Were the roofs higher or more sharply angled, that comforting effect would be lost.

Some people comment that the roofs of Jeju houses resemble the hills, known as *oreum* in the local dialect, that are spread across the island. They may indeed seem similar when viewed together, the *oreum* in the distance and the low, thatched roofs. However, an *oreum* is a very steep slope. In

The traditional roofs of Jeju houses have a very gentle incline, an adaptation to the strong wind on the island. (Seongeup Folk Village, Jeju-do. Jan., 2006)

that regard, there is a huge difference between an *oreum* and the thatched roofs on Jeju houses. If the roofs were as steep as the hills, the view from the airplane would seem much less peaceful.

The roofs of Jeju houses have a very gentle incline. Despite the fact that Jeju receives more precipitation than other regions, the traditional roof design is an adaptation not to rain, but to the strong wind.

A house of the same size, but with a sharply inclined roof, would look much bigger and more imposing, and feel more grand. If you visit the west coast area in the middle regions of the Korean mainland, you'll find that every house has a steep roof. At first sight, such houses seem quite large. But once you are inside of one, you quickly realize that it really isn't much larger than a comparable, but low-roofed house on Jejudo Island.

There are no crests on Jeju roofs. This is also a way to prevent the roof from being damaged by strong winds. Even on the mainland, roof crests

Modern houses on Jeju mostly have hipped roofs to minimize exposure to the wind. Stone-wall windbreaks beside the houses are built up to the same height as the eaves also to prevent damages by strong winds. (Jeju, Jeju-do. Aug., 2007)

are not pronounced in the coastal areas, but become larger and more decorative the further inland you go.

On Jeju houses, the height of the eaves is reduced in order to minimize exposure to the wind. In one shoreline village with especially strong winds, the stone-wall windbreaks beside the houses are built up to the same height as the eaves. Modern houses on Jeju mostly have hipped roofs (as seen in the above photo). This style of roof is favored on all of Korea's islands, since hipped roofs minimize wind damage.

The design of Jeju roofs results from abiding the principles and adapting to the forces of nature, without concern for ostentation or a desire to show off to others. If the roofs were too steep, they wouldn't be able to withstand strong winds. Traditionally, Jeju roofs are not only gently sloped, but covered with thatch. The thatch is tightly bound with braided straw ropes. The rope used on the thatched roofs of houses in the western part of Jeju, where winds

Traditionally, Jeju roofs are not only gently sloped, but covered with thatch that is tightly bound with braided straw ropes. (Seongeup Folk Village, Jeju-do. Jan., 2006)

are especiallly strong, is thicker than in the eastern part. Also, the ropes are set closer together. To prevent being blown about in the wind, the rope is tightly bound to a bamboo rod that hangs below the edge of the eaves.

Regarding the traditional architecture of Korean houses, it is generally said that as you go further south, the design of the houses becomes more open. This theory appears plausible if you are comparing the houses in the Hamgyeong-do (the northernmost Provinces in North Korea bordering with Russia and China) with those on the south coast. But the minute you arrive on Jejudo Island, the theory becomes suspect. Houses on Jeju are anything but open.

To enter a house built in the traditional Jeju style, you have to pass through either an ollae or an *imun'gan* stone-wall windbreak. An *ollae* is the passage from the road or an alley to the front yard of a house. With some houses, the *ollae* is long enough to be mistaken for a small alley, especially because of the high stone wall on either side. And there are some houses where a tall, dense windbreak thicket has been planted behind this stone wall. Because of the stone wall, the actual house cannot be seen from the road.

Some people say that the *ollae* walls were designed to block evil spirits from entering the house. However, in connection with the climate, *ollae* prevented strong winds from blowing directly into the house. Generally, these stone-wall windbreaks are built in a smooth curve. But even when they are built along a straight line, the *ollae* never meet with the hallway of the house at a right angle.

In front of an *ollae* wall, a *jeongnang* was always kept to show whether someone was home or not. A *jeongnang* is made up of three wooden rods balanced on top of two low stone columns, a low wooden fence-gate somewhat resembling a steeplechase jump. By placing the rods in different configurations, the owner of the house could inform potential visitors that

An *ollae* is the passage from the road or an alley to the front yard of a house on Jejudo Island. (Jeju, Jeju-do. May, 1994)

he had left the house, and how far he was planning to go. Such was the tradition. However, even back in the 1960s, when I was a child, that custom of signalling had almost vanished.

If a house was so close to the sea that there wasn't enough space to build the stone-wall passage, people built an *imun'gan*, which was also called *meonmun'gan* (literally, Distant Gate). Not only by the sea, but also in villages halfway up the mountain, where space is limited, *imun'gan* windbreaks were used instead of *ollae*. An *imun'gan* is not that different from the *daemun* (literally, Big or Front Gate) that was used in other parts of Korea. Therefore it is incorrect to say that there were no front gates on Jeju's traditional houses. Among the villages along the northern coast of Jeju, in those that used to be densely populated, there were many houses that had an *imun'gan*. This was also a response to the strong winter wind. Some people suggest that these distant-gate windbreaks were built in preparation for possible invasions by foreign forces. However, not all

An *imun'gan*, literally, Distant Gate, is the front gate of houses that did not have enough space to build the stone-wall passage, *ollae*, on Jejudo Island. (Seongeup Folk Village, Jeju-do. Jan., 2006)

coastal areas have distant-gate windbreaks. They are not common in the coast south of Hallasan, where the wind is weak. So, it is clear that the intended function of such walls was to block strong winds. An *imun'gan* is connected with with a high stone wall, and it is enough to block high winds from blowing into the main building.

On a day that the northwesterly winter monsoon was blowing fiercely, it was hard to see the inside of a traditional Jeju house, even if you had come in through the *ollae* passage or through the *imun'gan* gate and were standing in the front yard. This was due to the *pungchae* (a kind of pentroof), which is installed on the facade of a house and hung under the roof, to at once fasten the roof to the house and also act as an awning.

On very hot days, the awning support is lifted to block the sunlight; on stormy days, the screen is lowered to prevent the wind, rain, or snow from blowing into the house. Jeju's *pungchae* are similar to the *udegi* found on Ulleungdo Island, and the *kkadaegi* found on the mainland, in the Jeolla-

A *pungchae* awning support is lifted to block the sunlight on hot days in the summer. (Seongeup Folk Village, Jeju-do. May, 2006)

do.

Jeju's traditional houses generally have two wings, or more precisely, they have an extra row of rooms attached to the standard layout. Sometimes, in fact, there are two extra rows. Houses with an extra row or two of rooms are also built in parts of Korea with exceptionally cold winters, such as the Taebaeksan area. The extra thermal insulation provided by the extra layers of wall and air-space would be the obvious explanation. So naturally enough, in places where winters aren't so cold, or where the summers are muggy, there are no extra rows of rooms added to the standard floor-plan. The marked exception is Jeju, where the traditional houses had an extra row of rooms, despite the fact that the island has the highest average winter temperature in Korea. The explanation is to be found in Jeju's notorious winter wind.

The extra row of rooms served as an additional barrier to the cold north-

A double set of doors provided extra insulation from the cold northwesterly wind. (Seongeup Folk Village, Jeju-do. July, 2006)

westerly winter monsoon, keeping the house warm in winter. The presence of windbreaks certainly helped, but where windbreaks weren't sufficient, extra empty spaces were included in the construction of the house itself. Such spaces included a *nang'gan*, in front of each room, and a additional space at the back. In both cases, the function was to provide an extra, intermediate space, so the wind would not blow directly into the living area.

The doors formed another barrier to the wind. Having a double set of doors provides extra insulation by trapping a layer of air between them. This principle is embodied in the windows and balcony-doors of most modern houses in Korea, where two metal-framed, glass panes are mounted on horizontallly-sliding tracks, spaced slightly apart, maintaining a layer of air between them. On Jeju, the traditional houses achieved the same effect with a pair of doors made of of rice-paper glued to wooden

One can hardly see the houses that are surrounded by high stone walls on Jejudo Island. On the right are wind-shaped trees affected by the northwesterly winter monsoon. (Seogwipo, Jeju-do. May, 2006)

lattice frames.

Every now and then, one comes across Koreans from other Provinces who think that Jeju natives are cold and unfriendly. If there's any truth at all to this stereotype, it would be the result of the cold northwesterly wind. The bitterness of the wind on Jeju is hard to convey to anyone who hasn't actually felt it. For natives of the island, life was a constant struggle with that wind. That struggle may have engendered a closed, defensive attitude. The wind has entered deep into their lives.

Once, a person from Seoul visited Jeju and asked a passing student for street directions. The student answered very politely. But the Seoulite couldn't quite understand the answer, so he asked again. The student politely explained in more detail. At that, the visitor cried out, "Does everyone on Jeju talk so fast?" and gave up his inquiry.

Something similar happened on Ulleungdo. During a field trip, I saw a

An *u'ttal* is a windbreak, a high fence made of dry sticks or grass, that protected houses from strong winds on Ulleungdo Island. (Ulleung, Gyeongsangbuk-do. May, 2005)

perculiar kind of fence, so I asked an elderly woman what it was called. She gave a very short answer, unintelligible to me. So I asked again. Still not understanding her answer, I asked her to repeat it. After several more unsuccessful volleys, she became annoyed: "Why on earth can't you understand that word?" Later, when I looked it up, I discovered that her reply had been *u'ttal*. It was a shortened version of the standard Korean word for fence, *ultari*. But the woman's pronunciation of the word was so quick that it sounded like only one syllable. Therefore, it was incomprehensible to an outsider.

The dialect of the island areas tends to be coarse. Moreover, the words are shortened. And the people pronounce those truncated words in a very rapid stream. Accustomed to speaking this dialect, islanders often find it hard to adapt their speech in order to make themselves understood when they move to metropolitan areas on the mainland. The rapid-fire delivery and truncation of words that characterize island dialects is largely the result of climate. On most islands, the wind is so strong, and the wind-chill so cold, that people develop the habit of speaking quickly. Moreover, since the strong wind makes it hard to hear final sounds, the last syllable of words is eventually dropped.

The *kkadaegi* windbreaks of the west coast

Standing by the sea on the west coast, one feels the full force of the Jeju wind. When the northwesterly winter monsoon is raging, blasts of cold air pierce right through your skin, making it hard to stand still long enough to snap photos of the beach. Although I don't generally like hats, I developed the habit of wearing one on my frequent field investigations along the west coast. Sometimes I even find myself wishing I had ear-flaps. Standing out in the freezing cold wind, it feels as though my ears are about to fall off.

I once surveyed several areas, including Gimje and Jeongeup, to study the structure of houses on the west coast of the mainland. Although the

A map of villages in Gwanghwal, Gimje plains, Jeollabuk-do Since the road runs from northwest to southeast, the northwesterly winter monsoon whips right through.

cold weather made the work fairly demanding, I realized that winter field surveys are much more effective. Only when one looks around a house when the winter wind is blowing can one clearly understand how Koreans in the past provided against the wind. This is most starkly evident for houses built right by the shore. Gwanghwal-myeon, in Gimje, Jeollabuk-do is a good example of such a place.

Standing on the road that cuts through Gwanghwal-myeon in midwinter, you experience the reality of bitter wind. And since that road runs from northwest to southeast, the northwesterly winter monsoon whips right through. Things may change once the much-disputed Saemangeum land-reclamation project is completed, but as of now, both Gwanghwal-myeon, which is almost surrounded by the sea, and Jinbong-myeon, north of Gwanghal, are both renowned areas for cold winter winds. But even that wind seems warm after you've heard what the winds were like when farmers first moved to the area.

Gwanghwal-myeon is on land that was freshly created from the sea by a

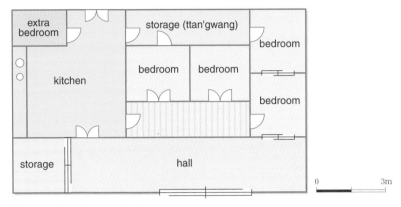

The floor plan of a house in Gwanghwal

reclamation project conducted by the Japanese Imperial Government, when Korea was under Japanese rule. When farmers from Japan were settled on Pyeonghwachon (literally, Peace Village), another area of reclaimed land in Gunsan, just north of Gwanghwal-myeon, Korean farmers were relocated to Gwanghwal, starting in 1933. When the first of the relocated Korean farmers arrived, they had no houses. All they had was some space to lie down, a bit of flat ground where they could boil a pot of barley, and a barn. It would have been difficult to escape the icy northwesterly wind. Moreover, Korea experienced extremely cold weather during that period. Those first residents of Gwanghwal must have felt like the ethnic Koreans in Russia, whom Stalin deported from coastal Siberia and relocated to Central Asia. So the farmers newly arrived to Gwanghwal devised a way to survive. Just as the ethnic Koreans in Central Asia excavated mud holes, the first settlers of Gwanghwal set up *kkadaegi*: windbreak fences made of straw. The bitter, salty wind raged across the reclaimed land, but the land was their hope for the future, and they didn't give up. As the settler families grew larger, the structure of the houses changed slightly. Rough, shed-like structures gave way to solidly constructed rooms, and the area between the *kkadaegi* windbreaks and those rooms was floored over with wood to form a halway. Despite the upgrades, parts of the original windbreak fences around the houses were left.

Although I had frequently traveled to the Jeolla-do, and had visited the fields of Mangyeong and Gimje, it was many years before I learned about the *kkadaegi* windbreak, and then it was by mere coincidence. I told an old local acquaintance that I was doing some research on the design of houses in that area, and he immediately asked me whether I was familiar with the *kkadaegi*. That piqued my curiosity to such an extent that I stayed up that entire night, waiting for dawn to break. But when I inspected the sites, there were no *kkadaegi* to be found. For years after that, for all my interest in the structure of Jeolla houses, I never saw a *kkadaegi* in its original form.

The *kkadaegi* windbreaks seem to have been made for various purposes, but the primary function was to block the wind. A *kkadaegi* is a wall made

Vestiges of *kkadaegi* can be seen anywhere in the Jeolla-do, both along the coast and inland, in the form of awnings that prevent rain or snow from blowing into the house. (Naju, Jeollanam-do. Feb., 2008)

of straw that follows the edge of an awning that is extended from the eaves when the cold winds start to blow. This straw-wall windbreak is sufficient to block the cold northwesterly wind and the snowstorms that accompany it. This helps to keep the house warm. When the cold northwesterly winds diminish and the warm spring weather is about to arrive, the straw-wall windbreak is removed. The fact that these *kkadaegi* straw walls are installed every year in late autumn, and removed in the following spring, is what makes them different from the udaegi of Ulleungdo. Vestiges of *kkadaegi* can be seen anywhere in the Jeolla-do, both along the coast and inland.

The straw-wall windbreaks were hung not only at the front and the back of a house, but also on the sides. The space between the side-*kkadaegi* and the house was used as a storage place for farm tools. As a result, the straw walls on the side are in place almost the entire year, and look like an integral part of the house structure. Also the straw wall in the back was

Clear vinyl sheets act like a greenhouse, allowing the solar energy into the hallway, but preventing the heat from escaping. Vinyl walls are installed in place of *kkadaegi* windbreaks from late autumn at villages in Jeolla-do. (Wanju, Jeollanam-do. Oct., 2006)

incorporated into the structure of the house, in case the house needed to be enlarged. At Gwanghwal, such space is called *ttan'gwang* (literally, 'other attic').

Today, these walls at the back and on each side of a house are no longer made of straw; they are proper walls of brick and cement that seem to be an integral part of the house. The front windbreak, however, is generally not a permanent structure. These are usually made of vinyl, and are only installed in the winter. But this construction is very effective: even in midwinter, with a northwesterly wind raging outside, it is very warm in the space between the windbreak and the house. The clear vinyl acts like a greenhouse, allowing the solar energy into the hallway, but preventing the heat from escaping.

By the shore or on islands, people put up an outer gate instead of a *kkadaegi*. When I came upon a house with two gates, it was in the middle

A house on Gaeyado has two gates and the roof is tied to the ground with ropes to resist strong winds. (Gunsan, Jeollabuk-do. July, 1998)

of the summer, the hottest time of year. Even without the extra gate, it would have been very hot. It led me to assume that the winter wind in that area must be very extreme. For the elderly owners of the house, it may be easier to just endure the extra heat in summer than to remove the outer gate at the end of winter, and put them back up again each year.

Most of the roofs of the houses that have an outer gate are tied with ropes. They are tied firmly to prevent the whole roof from blowing away. These are strong nylon ropes of the sort used for towing cars. Houses with these low, gently sloped, hipped roofs make a small village look even more modest. Still, these houses are surrounded by high stone walls. And to the number, such houses have high chimneys. High chimneys are a necessity in places with a lot of wind. Otherwise, instead of exiting smoothly and venting outside, smoke would get backed up into the fireplace.

In no matter what region of Korea, wherever winds are strong, houses

High chimneys are a necessity in places with a lot of wind. Otherwise, instead of exiting smoothly and venting outside, smoke would get backed up into the fireplace. (Yesan, Chungcheongnam-do. Jan., 1991)

have high chimneys. This is readily observed both on islands and along the coast. Among the places I have surveyed, chimneys are especially high on Ganghwado, and in Yesan, on the Taeanbando, and along the west coast of the Jeolla-do. On the other hand, in the inland areas, where the wind is weak, the chimneys are quite low.

Visiting the Cheongpung Cultural Property Complex, where old houses have been preserved for historical interest, it is hard to notice the chimneys on the houses. That is because the chimneys are quite low, and are located at the back of the house. On Jejudo Island, where the wind is very strong, houses have no chimneys. The reason isn't clear, but it may have something to do with the fact that there are separate fires used for heating the house and for cooking. Unlike the lifestyle on the mainland, in Jeju houses, the firewood used in the kitchen stove for cooking is not the same one burned in the fireplace for heating the floors of the house. So there being no ondol, there was no chimney.

15

Adapting to a snowy climate

The houses in the Nari Basin on Ulleungdo Island have very steep roofs to prevent the accumulation of snow. A windbreak made from eulalia grass provides a passage to the house when the snow piles up to the eaves in midwinter. The houses in Gangwon-do, east of the Taebaek Mountains, have a similar windbreak, and most of the domestic activities are centered around the kitchen.

P uppies and children simply love the snow, but older people may take a different attitude. Those of us in between risk being seen as senior citizens if we make unappreciative remarks about snow. But it's an inevitable feature of winter, and everyone recognizes the beauty of the falling white flakes. The season's first snow may be one of the most eagerly awaited and heartily greeted of meteorological moments. Many people even take time out from their routine activities in order to greet it. But not too long after the joyous celebration of the first flakes, the excitement and novelty starts to wear thin. Especially among drivers of automobiles, snow is often less than fully welcomed.

When I was in elementary school, I was always delighted to find that snow had fallen during the night, even if it had cut off our house from our neighbors. For children who didn't have many personal toys to play with, the snow itself was a marvelous plaything. Even plowing the snow in front

A village completely covered with accumulated snow. (Gochang, Jeollabuk-do. Dec., 2005)

Since most residents in the countryside are old, heavy snowfall is a source of anxiety. (Jeongseon, Gangwon-do. Jan., 2008)

of the house was exciting. Nowadays, however, since most residents in the countryside are old, heavy snowfall is just a source of anxiety. For people in their eighties, clearing snow from in front of their house is no pleasure. In villages where there are still enough young farmers, all sorts of farm equipment is mobilized for the task.

As for us children, we preferred snow covered fields to the bare ground for flying our kites. As a further delight, a covering of snow invited us to slide down the slopes on sleds. And back in the 1960s, when I was a child, there were hardly any cars on the streets, so the snow turned any inclined road into a playground. The sight of snow starting to melt in the fields was an announcement that winter was nearly over, and so was winter vacation. So we felt wistful as we watched the snow gradually disappear.

Years later, the snow I saw while doing my military service was strikingly different from the snow I remembered from my childhood especially the snow at the Air Force base where I was serving. Since snow on the runway made it difficult for planes to take off or land, men of all ranks were required to stay on the base when snow was forecast. So everyone paid close attention to the meteorological team's forecasts. The team would be in a terrible position if a forecast for snow turned out to be wrong. But even when snow came as forecast, it was no cause for joy. As soon as the snow stopped, we all had to dash out to clear the snow from the runway. At the first hint of snow, all eyes were on the meteorological team. On the other hand, off-base and off-duty, the sight of snow delighted me. Ever since, I've recognized the two aspects of snow: the natural beauty we so welcome, and the many inconveniences we dread.

The *udegi* on Ulleungdo Island

Despite my professional interest as a geographer, it was a long time before I first set foot on Ulleungdo Island. In conversations and lectures about introduction to geography, among the first places I used to mention was Ulleung. I gave the impression of great familiarity with the place, which in fact I'd never visited. I used to regale students with stories about the prodiguous amount of snow on this island I'd never actually seen. Deep down, I always felt a bit uncomfortable about this – the guilt of false representation.

Then finally one winter, with great expectations, I boarded a ferry for Ulleungdo Island. People asked why I'd chosen to make the visit in winter, but to me it was the apropriate season: to experience Ulleung, you must see the snow. The three-hour trip took longer than scheduled, but after four hours at sea, the island finally came into view below a hazy sky. My eagerness to explore mounted as we approached. If what I'd been telling students in my lectures was true, the weather that day fit the conditions for a heavy snowfall. As if granting my wishes, when the ferry landed, snow

Snow covered Nari Basin on Ulleungdo Island, one of the parts that particularly receives a great deal of snow. (Ulleung, Gyeongsangbuk-do. Jan., 2003)

was falling on Ulleung, and the island was covered by dark clouds.

But that was all the snow I saw on the island. I ended up being stuck there for the next ten days, but it didn't snow again. The northwesterly winter monsoon was blowing, with a wind and wave keeping all the boats tied at the piers. Yet for some reason, it didn't snow. As though the winds and the clouds were in a conspiracy, the amount of snow falling on Ulleungdo Island had apparently been in decline since the 1980s, when the issue of global warming began to be discussed as an international concern. Indeed, snowfall remained infrequent on the island for almost two decades before it started to pick up again from the winter of 2000.

The weather station of Ulleungdo Island is to the south of Seonginbong Peak, so the measurements do not reflect the situation in the Nari Basin or the area of Albong, two parts of the island that receive a great deal of snow. Ulleungdo is not a large island – at about 72.9 km^2, it ranks eighth in size among Korea's islands – but within that small area, there are remarkable local differences in the amount of snowfall. At Dodong Port, where many tourists visit, there is never a trace of snow. But in the Nari and Albong Basin, quite commonly, there is so much snow that access is limited to vehicles with four-wheel drive.

Three years after that trip, I finally got an opportunity to fully experience Ulleungdo Island's famous snow. Once again, winter weather made my visit difficult. It was only after a week of patiently waiting in the mainland port of Pohang that I was finally able to board the ferry for Ulleungdo. Wind and wave warnings had kept all boats in port that long. But the week's delay gave me a better chance of seeing snow on the island, and there were news reports of snow in the Albong and the Nari Basin: exactly what I'd been waiting for. When I called to the local resident in Nari Basin contact from a village called Cheonbu on the northern coast of Ulleungdo Island, and asked about getting into the Nari Basin, I was told it would be difficult for our group to get there in the vehicle we had. He offered to come and pick us up. After some wait, a four-wheel-drive truck of the sort used on farms arrived. As soon as we left Cheonbu, we entered another world. Even though we were riding in a solid, capable vehicle, I couldn't help feeling nervous. The road was so steeply sloped that it was hard to stand straight, and above that was a mountaintop covered with a deep, deep layer of snow.

It was an arduous, difficult trip, but well worth it: the snowy landscape we saw fully lived up to its reputation. And the sheer volume of snow made it easy to understand why the island's houses had come to be built as they were. With the snow reaching up to the eaves, it was hard to tell, when

Hiking on the way from Nari Basin towards Albong peak There is still a lot of snow on the way to Albong although it's supposed to be spring. (Ulleung, Gyeongsangbuk-do. April 1, 2003)

viewed from above, whether there was even a house. And after that winter, I revisited the Albong Basin in early April, when spring had arrived. There were still snowfalls accumulating ankle-deep. There was still snow beneath the trees on the surrounding mountains. According to the local residents, it was common for snow to remain piled higher than the eaves all winter.

The houses in the Nari and the Albong Basin, unlike those in other parts of Ulleungdo Island, have steep roofs, as sharply inclined as the surrounding mountain peaks. Because of the steep tilt, snow slides off the roof instead of accumulating on top. Otherwise, the weight of accumulated snow would be more than the structure could bear. The houses along the coast of Ulleungdo, on the other hand, have gradually sloping roofs, clear evidence that little snow falls there. Even when it does snow, it soon melts, rather than piling up on the roofs. Thus, houses along the coast appear to

Houses by the coast of Ulleungdo Island The houses along the coast of Ulleung have gradually sloping roofs, clear evidence that little snow falls there, and a small yard with an *uttal* windbreak. (Ulleung, Gyeongsangbuk-do. May, 2006)

be designed with strong winds, rather than heavy snow in mind.

When Koreans think of the houses on Ulleungdo Island, *udegi* is the first word that comes to mind, thanks to what they learned form their middle- and high school textbooks. But not that many Koreans are aware of the Jeolla Provinces' equivalent: *kkadaegi.* Yet barely ten thousand people live on Ulleung, while the population of the Jeolla-do, including the city of Gwangju, is close to five million. The fact that most Koreans have heard of *udegi* but not *kkadaegi* shows the strong influence of textbooks in Korea.

Udegi and *kkadaegi* are similar in shape and related in function. *Udegi* is a sort of a wall made of thatched eulalia grass, and *kkadaegi* is a similar structure made of rice straw. But while an *udegi* stays in place throughout the year, a *kkadaegi* is only put up during the winter. And while the main purpose of a *kkadaegi* is to block the cold winter wind, an *udegi* is meant to

***Udegi*, thatched eulalia grass wall, and *jjukdam* passage** The space between the *udegi* and the house wall is called *jjukdam*, and it is used as a passage. The *udegi* of houses by the coast also play an important role as windbreaks. On the left is the house wall, and on the right, the eulalia grass wall. (Ulleung, Gyeongsangbuk-do. Jan., 2003)

prevent snow from blowing into the house. As snow frequently piles up to the eaves of a house in midwinter, the snow could block the front door, making it impossible to enter or exit the house. So, an *udegi* is to prevent snow from piling up at the entrance and thus secure access to the house. The space between the *udegi* and the house wall is called *jjukdam*, and it is used as a passage. The *udegi* of houses by the coast also play an important role as windbreaks.

Villages in the highlands can be cut off from communication by a snowstorm. This kind of isolation occurs even today. Tourists occasionally visit villages in the Nari Basin even in winter, but in highland villages, heavy snows cut off communication with the outside world completely. At such times the small, snow-isolated villages are reconnected by cable-ways.

These cable-ways have been used on Ulleungdo Island for a long time to deliver supplies to small, snow-isolated villages. (Ulleung, Gyeongsangbuk-do. May, 2007)

These cable-ways have been used on Ulleungdo Island for a long time to deliver supplies to isolated villages. Nowadays they are used to deliver mountain herbs, a local specialty product. The fact that the whole island is a steep slope led to the early development of these cable-ways.

The *tteureok* of the Yeongdong region

Another part of Korea with a snowy climate is the Yeongdong region: the part of Gangwon-do east of the Taebaek Mountains. One drizzling autumn day I was in that region, surveying a place called Geumgwangpyeong, near Gangneung. During some spare time I stopped by a village upstream of a reservoir at the edge of Geumgwangpyeong. The village comprised about a dozen households, and I was intent on examining one of the houses in detail. The owner had spread red chili peppers on a mat to dry, even though it was raining. The space where the mat was spread was no doubt the *udegi* and *jjukdam*: the passage between the windbreak wall and the house. However, at that time, I had not seen an udegi yet and I wasn't even aware of the existence of *jjukdam*, so I failed to notice this interesting phenomenon.

Since that day, I have often visited Geumgwangpyeong. Going there in winter, the particular space between the wall and the house that I'd seen at other seasons of the year struck me afresh. Eventually, I became curious about the purpose of the space, and I ended up investigating how houses were built in areas where it snows a lot.

When surveying a house in the Yeongdong region in the winter, when the owner came out to greet me, nine times out of ten it would be through the kitchen door. This would be surprising to people from outside the region, who would naturally expect the owner to appear through the big doors at the center of the house. When entering the house, one first walked

Tteureok, the foundation platform of houses east of the Taebaek Mountains, 30 to 90 cm high, is also used as a passage or a verandah to dry vegetables, such as red chili peppers. (Gangneung, Gangwon-do. Jan., 2001)

Tteureok platform serves as a passage between rooms and also as the space beside a house to keep firewood piled up for the winter. In addition to serving as fuel, the piled logs also block the snow from blowing directly against the house. (Gangneung, Gangwon-do. Jan., 2001)

Tteureok embody highly adaptive responses to the many challenges that the heavy snowfalls pose to the residents of Yeongdong. First, the extent of a *tteureok* corresponds to the edge of the eaves in a manner that allows snow falling from the roof to land outside the *tteureok*. Therefore, no matter how much it snows, within the *tteureok* it will be mostly dry. (Gangneung, Gangwon-do. Feb., 2002)

into the kitchen, then stepped through a small door that led to the living room. Not just the kitchen door, but the entire enclosed area of the house was built on a raised platform, and what caught my attention was how high above the ground that platform was built. Although it varied from house to house, the platform was always significantly elevated: in many houses it was even too high for a person to mount directly without the aid of a step. This foundation platform is called a *tteureok*.

In the areas east of the Taebaek Mountains, every house has a *tteureok*, and their heights range from 30 to 90 cm above ground level. In many houses a step is built into the *tteureok* for easier access to the kitchen door. In coastal areas, the *tteureok* are relatively low, but as you move to higher altitudes, you find the *tteureok* are built further off the ground. The reason is simple: there is more snowfall at higher altitudes. The platform serves as a passage between rooms. In the space beside a house, firewood is kept piled up for the winter. In addition to serving as fuel, the piled logs also block the snow from blowing directly against the house.

Tteureok embody highly adaptive responses to the many challenges that the heavy snowfalls pose to the residents of Yeongdong. First, the extent of a *tteureok* corresponds to the edge of the eaves in a manner that allows snow falling from the roof to land outside the *tteureok*. Therefore, no matter how much it snows, within the *tteureok* it will be mostly dry.

Today, as nearly all houses use boilers for heating, more and more people have installed aluminum-frame doors in front of the *tteureok* to increase thermal efficiency. There was no use for these sash doors when people burned firewood. Although the material is different, these glass sash doors are functionally similar to *udegi* windbreaks.

In the Yeongdong area, there is no need for windbreaks like the *udegi* of Ulleungdo Island or the *kkadaegi* found in the Jeolla-do, because the nature of the snowfall is different. The snows of Ulleung and Jeolla are driven by the firece northwesterly wind, while in Yeongdong, snows result from a

Today, as nearly all houses use boilers for heating, more and more people have installed aluminum-frame doors in front of the *tteureok* to increase thermal efficiency. (Samcheok, Gangwon-do. Jan., 2000)

migratory high, with much milder winds. So, in contrast to the wind-whipped snows of Ulleung and Jeolla, snowfalls in Yeongdong, however heavy, and however great the accumulation, are quiet and gentle. So, there is no need to build a separate wall; all that is required is to elevate the living area above the snow covered ground. To that end, a *tteureok* serves admirably.

Once when I first stepped into the kitchen of a house, I turned to my companions and jokingly suggested a game of football. The kitchen seemed almost big enough. The kitchens I remembered from the houses on Jejudo Island were fairly large, but they in no way prepared me for the vast kitchens I saw in this region of the east coast. I soon learned the reason: when there is a lot of snow piled up, the kitchen becomes the household's main living space. Therefore, the kitchens were spacious, and were conveniently connected to all the rooms of the house. Even the cowshed was connected with the kitchen. It was another unfamiliar sight

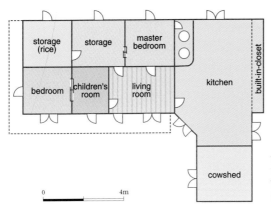

storage (rice) · storage · master bedroom · kitchen · built-in-closet

bedroom · children's room · living room

cowshed

0   4m

The floor plan of a house in Yeongdong region, east of the Taebaek Mountains (Yangyang, Gangwon-do)

for someone who grew up on Jeju, where the stable is very far away from the kitchen. However, thinking of heavy snow piled up in midwinter, it seemed natural that the cowshed be close to the kitchen. After boiling the cow's chaff in a pot on the kitchen fire, one could pour some straight into the trough. Not all parts of the house are directly connected with the outside, but within the house, all parts are internally connected. The pattern of concentrating all household functions in or near the kitchen is common to houses located in the very snowy parts of the country. It contrasts sharply with houses in areas of little snowfall, where household functions take place in separate spaces, scattered throughout a compound.

In mountainous areas of Korea with especially heavy snowfall, some houses have an exit door in the ceiling of the kitchen. When the house is virtually buried in accumulated snow, making it difficult to exit, the owner climbs a ladder up to the kitchen's ceiling door, and walks out onto the roof of the house. From there it is possible to sweep the snow from the top of the house, or to communicate with neighbors standing atop the roofs of their own snowbound homes.

16

Meeting the challenges of hot and cold weather

On the three hottest summer days each year, when Koreans seek food to boost their health, they turn to ginseng-chicken soup (*samgyetang*). This hot soup helps people regain energy when their bodies are weary from weeks of the rainy season, followed by the muggy heat. Because of the hot temperature, Korean cuisine relies on spices as preservatives, to prevent foods from going bad quickly. There are vegetables pickled in bean paste or red chili paste, and a variety of fish-based sauces. On the other hand, to meet the challenges of the severe winter cold, a floor-heating system, called *ondol* (literally: warm stone) was developed, and Koreans developed preserved vegetables, such as winter kimchi, made by salting and fermenting Korean cabbages and other sliced vegetables.

The Korean climate is characterized by extremes: sweltering heat in summer; numbing chill in winter. Most countries in the temperate zone also have four seasons, but in few places are they as sharply differentiated as in Korea.

What would Korea be like without its hot summer? The difference would be catastrophic. Koreans may find it difficult to bear the scorching heat of summer, but it is essential to the country's culture as it has developed. It is a rice farming culture; Koreans eat food made from rice from the very first meal on New Year's Day. This is only possible thanks to the intense heat of summer. Without their hot summers, Koreans would not be able to eat domestically grown rice at the start of the year.

Early at dawn one late-autumn day in 1980, I was driving along National Road 36 from Bonghwa to Uljin, in Gyeongsangbuk-do. I had been driving on the unpaved road for a considerable time when the sun rose. Glancing out the window, I saw an extraordinary sight that made me doubt that I had fully awakened from a dreamy sleep: a field of

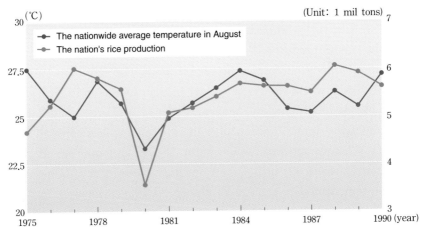

The nationwide average temperature in August and the nation's rice production
Without the sweltering heat of August, the rice plants had been deprived of their full growing season.

unharvested rice plants. They were white with the frost that had arrived before they had fully ripened; the very frost that had made the farmer give up on the harvest. A truly heart-breaking sight.

What had happened here? Rice plants are supposed to bow their heads when they are ripe, but here they were, standing upright in November. The summer had been cool that year: the nationwide average temperature in August about 3°C lower than usual. And those cool August temperatures were to blame for the rice plants' standing so rigid and unripened in November. Without the sweltering heat of August, the rice plants had been deprived of their full growing season. Absent were the usual autumn news reports of golden fields and abundant harvests; it was a decidedly lean year. The nation's rice production in 1980 barely reached sixty percent of the previous year's. If such poor harvests continue for several years, a nation can be emperiled. The mid-nineteenth-century Irish Potato Famine stands out as an extreme example. Yes, Korea needs its hot summers; they keep the agricultural cycle running smoothly.

And what about those cold winters? What would Korea be like without them? Again, the result would be a disastrous disruption of the agricultural cycle. For Koreans, the words Daegu and Gyeongsan used to be synonymous with the famous product of those regions: apples. But that instant mental association no longer reflects reality. Apple-production has moved further north, to areas where winters are colder than in Daegu and Gyeongsan. Places like Andong, Bonghwa, Yeongju, Chungju, and even Yeongwol are now supplying apples. Successful fruit farming requires suitably cold winters. When winters become too warm, an area no longer provides the conditions farmers rely on for their orchards. Patterns of farming must change to accomodate changes in climate. The phenomenon extends beyond apples.

It used to be that the only place in Korea where tangerine oranges were cultivated was on Jejudo Island. But it is no longer news that this sub-

For Koreans, the words Daegu and Gyeongsan used to be synonymous with the famous product of those regions: apples. But now apple-production has moved further north, to areas where winters are colder. Facilities are mounted throughout the orchard to prevent late-frost damages. (Yeongwol, Gangwon-do. Sept., 2008)

tropical fruit is also being produced in orchards further north, in areas along the south coast of the Korean mainland. Unfortunately, those tangerine orange farmers along the south coast have suffered as much as the ones on Jeju with the opening of the Korean market to imported agricultural goods. Indeed, farmers in Goheung County have recently been chopping down their tangerine orange trees.

The intense cold of winter serves another vital function for farmers: it eradicates diseases and pests from the fields. Without that winter cold, farmers would have great difficulty growing healthy plants during the following summer.

The traditional dietary wisdom of *samgyetang* and winter kimchi

Nowadays, Koreans can chose from an extensive menu of chicken dishes at any time of the year. No modern Korean would regard a piece of boiled chicken leg as medicinal tonic for the elderly. Yet as a child, I used to carry a bowl containing a single piece of dark chicken-meat to the house of an elderly neighbor once a year. When I was a bit older, I learned that the day of that annual delivery was one of the hottest days of the year. Wealthy families would kill three chickens a year, one for each of the three days traditionally regarded as hottest. But most families could not afford such luxury. They had to be satisfied with one chicken a year, and even that was considered good fortune. Memories like that drive home the magnitude of the changes we've lived through. Still, Koreans recognize the tonic benefits of hot chicken soup. On those three hot summer days they serve a ginseng-chicken soup called *samgyetang*. However, instead of dividing a single chicken among an entire family, the standard portion now is one chicken per person.

The three traditional 'hot' days are seen as requiring Koreans to consume highly nutritious foods like *samgyetang*. First, around July 15, is the 'early hot day' near the end of the rainy season, when people feel the need for an energy boost. Farmers, in particular, needed to revitalize themselves in order to go out to weed the rice fields and vegetable beds. Next comes the 'middle hot day', around July 25, at the very end of the rainy season. The special nutritional boost is intended as preparation for the upcoming sweltering heat. And the final hot day, around August 14, comes towards the end of the summer hot season, when people felt a need to replenish the energy that had been depleted by the heat.

In my childhood, until the first summer vegetables came up, there was only one dish placed on the table to accompany the rice. It was a young garlic stalk, cut in two-inch sections, and preserved in soy sauce. On Jejudo Island, people call this dish *manong jisi*, literally, garlic-stem pickles. On the mainland, people have a similar dish, called *maneul jang'ajji*, (literally, garlic pickles), made with a fully grown garlic stalk. It seems that the islanders do not have the leisure to wait for the garlic to grow to maturity. On Jeju, people prepared the garlic-stem pickles in spring, and ate them until midsummer. As time passed, white fungus would grow on the surface, but, surprisingly, it caused no stomach-upset when eaten. A soy-sauce pickle made with vinegar was a luxury beyond what we could hope for. As school children, our lunch-boxes contained cooked barley and two or three garlic-stem pickles. That sufficed. Even the children of the neighborhood's so-called wealthy families only had a few dried anchovies as an extra side-dish.

In any area where seasonal changes limit the growing season, you will find some culinary technique for preserving fresh vegetables to be eaten later in the year. Just as vegetables pickled in brine have long been a part of the Western diet, Koreans have a wide range of dishes, collectively known as *jang'ajji*, based on vegetables preserved in salt, soy sauce, or red chili paste. Nowadays, vegetables preserved in red chili paste are especially valued for their complex savor: the longer they've been aged, the more delicious and costly. The southern region of Korea is famous for its varied and delicious *jang'ajji*. For several centuries, Sunchang, in Jeollabuk-do, has been famous for its top quality red chili paste (*gochujang*). Recently the local government built a *gochujang* 'village' on the outskirts of Sunchang, and is vigorously promoting the *gochujang* and preserved vegetables produced there. The effect of this promotion, however, has been to degrade the reputation of Sunchang gochujang, which was originally based on the products of a downtown neighborhood.

Growing up, I was only familiar with one type of preserved vegetable: garlic pickles. But now I encounter all sorts of items: turnips, green chili pepper, deodeok root, Korean lettuce, persimmon, Korean melon, yellow corvina.... It seems that pretty much anything that people can eat can be preserved. Natives of Ulleungdo Island regard a dish called *san maneul jang'ajji* as a dinner-table essential. Made by pickling the leaves of

Sunchang County built a new village to promote the local specialty, red chili paste or *gochujang*. (Sunchang, Jeollabuk-do. July, 2008)

mountain garlic, it is known as a life-sustaining herb for helping the islanders survive the years of food shortage. The smell of the soy sauce, however, is a bit overpowering to people from the mainland.

Preserved fish or salted fish is also a very important ingredient in Korean food. It is a key element in making good kimchi. Koreans in all regions of the country keep a few kinds of salted fish in the refrigerator; no Korean table is complete without it. Different kinds of fish are caught in each region, so there is considerable variety in salted fish : more than a hundred kinds in all. The most commonly used salted fish are preserved shrimp from the west coast, preserved anchovy from the south coast, and preserved squid from the east coast. Other popular types, include salt-pickled pollack entrails, salted pollack roe, salted yellow corvina, and salted Pacific cutlassfish.

Fish caught in late spring are preserved in salt, and they are ready to eat

Ganggyeong in Chungcheongnam-do is now far from the sea, but it is renowned for its salted shrimp market that was established in the days when fishing boats came there. (Ganggyeong, Chungcheongnam-do. Feb., 2008)

from late summer or autumn. Many port areas along the west and south coasts are renowned for salted fish. Ganggyeong and Gwangcheon, in Chungcheongnam-do, are now far from the sea, but they still maintain the big preserved-fish markets that were established in the days when fishing boats came there.

Salted fish is the best side-dish for people who have lost their appetite in the hot, exhausting summer. Still today, if I go back to my hometown and there is salted damselfish on the table, my appetitie is stimulated to the point where I end up feasting. Preserved fish is more than just a way of keeping fish edible over a long period. In the hot summer, when people's vigor flags, side-dishes made from salted fish have a tonic, appetite-reviving function.

Preserved shrimp is one of the essential ingredients in making kimchi.

The best salted shrimp is that made from the June catch. The shrimp is at its best, and in addition, the summer heat helps the fermentation. Whenever I go on a survey to Gwangcheon, Chungcheongnam-do, I always stop by the salted shrimp stores. Gwangcheon was famous as a gold-mining town during the Japanese colonial period. Today it has gained new fame by using the former gold-mine pits as fermentation cellars for salted shrimp. Even people who are not particularly fond of salty food find their mouths watering at the sight of the creamy shrimp, and cannot resist the temptation to taste at least one. When I visit the shops with my students, I always feel embarrassed in front of the owner. My young students start jumping over each other in their eagerness to try one more sample of salted fish. There's no stopping them.

Jeju natives particularly like salted anchovy and damselfish. Salted anchovy is also made in June and ready to be eaten from July as an

Gwangcheon was famous as a gold-mining town during the Japanese colonial period. Today it has gained new fame by using the former gold-mine pits as fermentation cellars for salted shrimp. (Hongseong, Chungcheongnam-do. Oct., 2002)

important summer delicacy. The rich flesh made it the best side dish for farmers working out in the fields in the midsummer. Jeju people eat salted anchovy by wrapping it in a soy-bean leaf. In some parts of Gyeongsangbuk-do, people eat soy-bean leaves that they picked in the summer and preserved in soy sauce or soybean paste. That is also a sort of pickle. Koreans on the mainland use salted shrimp instead of salt when they cook pork, but Jeju people use salted anchovy. Chujado, the islands northwest of Jeju, is a popular place for anchovy sauce. One can taste salted anchovy not only on Jeju, but at any place near the south coast.

Ulleungdo Island is famous for squids to the extent that it could be called "Squid Island." Like the preserved fish anywhere else, the squid was an important source of protein, but on Ulleungdo, squid played that role year round. Fishermen start to make salted squid in October, and eat it all winter. And with the squid that are newly caught the following spring, they

Chujado, the islands northwest of Jeju, is a popular place for anchovy sauce. (Chujado, Jeju-do. April, 2008)

prepare salted squid in May and live on it through the summer. Salted squid is not only popular on Ulleungdo, but also along the east coast of the mainland: from Gyeongsang-do in the south up to the Hamgyeong-do North Korea.

Kimchi was traditionally the preserved food that got Koreans through the winter. Just as Koreans can eat chicken whenever they want to, nowadays Koreans can eat kimchi throughout the year. However, this only became possible fairly recently. In the 1960s, one rarely saw a dish of kimchi on the table in the summer. I think it was only after Korean cabbages started to be cultivated on highland farms in Gangwon-do that Koreans could eat kimchi even in the summer. Today Koreans take kimchi for granted. Maybe if they go abroad, they are reminded of how special it is. But as recently as thirty years ago, kimchi was almost the only thing Koreans had to eat with their rice in winter.

Each region makes a different type of kimchi, and the preparation occurs at a different time in the winter. When I stayed at a boarding house in Seoul, the host family was busy preparing winter kimchi in November. Seoulite had their winter kimchi to eat in November at the latest. But when I went home to Jejudo Island for winter vacation, there were no signs of kimchi preparation. That grand project wouldn't come until the end of the year, nearly an entire month later than in Seoul.

The difference between the kimchi making times in different regions is a function of climate differences. The winter climate in Korea shows a relatively wide range of regional temperatures. So in November, the Korea Meteorological Administration announces the proper time for making kimchi in each region. The best period for preparing winter kimchi is said to be when the daily minimum temperature remains below 0°C, and the daily average temperature falls below 4°C.

The higher the winter temperature, the faster the kimchi will ferment. Some people make their kimchi salty or add more spices to delay the

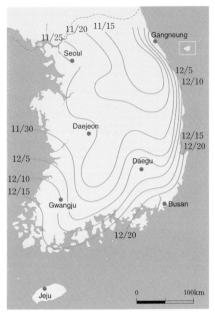

The winter kimchi front chart (Korea Meteorological Administration, 2007)
The winter climate in Korea shows a relatively wide range of regional temperatures. So in November, the Korea Meteorological Administration announces the proper time for making kimchi in each region: namely, mid-November in the mountain areas in Gangwon-do, and mid-December in the south-coast area.

fermenting process, so it is natural that kimchi is saltier and spicier the further south you go down the peninsula. For that reason, when I first visited the Jeolla-do, I became hesitant to point my chopsticks over towards the kimchi dish. The kimchi of that region was different from that was I used to. It had much more seasoning and fillings, and it was somehow not the same. But that saltiness and wide variety of seasonings is what makes Jeolla kimchi distinctive. Typically some sort of seafood is added, such as oysters, along with a generous portion of sun-dried red chili powder. Once one becomes used to the flavor of kimchi from the Jeolla region, it is hard to eat kimchi in other parts of Korea. When my roommate brought back some kimchi from his hometown Ulsan (Gyeongsangnam-do), it was also not to my taste. I was slicing into the kimchi when I encountered a piece of cutlassfish. I did not know till then that in Ulsan people add pieces of fish

– sometimes a whole fish – when they make kimchi. Back then I was dumbfounded, but now I enjoy the taste.

At first I had difficulty eating the kimchi at the house of my aunt, who was married to a man from the Chungcheong region: there was so much salted shrimp in it. But after living with her family for some two years, I gradually became used to that taste. And years later, when I married a woman from Gangwon-do, I once again had to adapt myself to a new kimchi taste. I could not dare tell my new bride back then, but her region's kimchi didn't seem to measure up to the kimchi I had become accustomed to at my aunt's. The kimchi of Gangwon-do, to the west of the Taebaek Mountains, seemed to have even less saleted fish than Seoul kimchi. Except for my native Jeju kimchi, Seoul kimchi best met my tastes. The amount of salt and red chili powder in Seoul kimchi seemed to be similar to that of Jeju.

Meanwhile, the Pyongan-do in North Korea are famous for winter turnip kimchi, *dongchimi*, which is not too salty, but slightly vinegary, and contains no red chili powder. When fairly ripe, dongchimi water is also drunk as a refreshing drink. And finally in the coastal Hamgyeong-do, a lot of seafood is included in the kimchi. The cabbage is not too salty, but plentiful amounts of other seasonings give it an appetite-stimulating taste.

There is also a nutrition-boosting health food for the winter. When winter was about to start, my mother was stuck in the kitchen for the entire day. She'd be stirring a big pot, spreading a sweet, delicious smell thoughot the house. She was making a special soup called *gol* on Jejudo. It was made of pheasant meat, procured with considerable effort, boiled with malt and plenty of water. My mother could not leave the kitchen for fear that the precious pheasant meat might stick to the bottom of the pot and burn. When she couldn't find a pheasant, she cooked not chicken, but Jejudo Island black pork. Her effort would culminate in the evening. What had been boiling all day in the huge iron pot would be shrunk to the size of one rice pot: pheasant syrup.

Pheasant syrup was kept hidden well above a wooden box on the top shelf of a built-in closet, and doled out in tiny portions throughout the winter. Being the oldest son may have entitled me to a privileged portion, but even that was only about one spoonful a day. But that sufficed: it may have been thanks to that high-protein syrup that I was always in good health during the winter. I do not know how people in the cities faced the challenge of winter, but in the countryside, most people had their own methods, such as pheasant syrup. When the syrup had nearly run out, the long, tiring winter would be nearly over.

Clothing and heating to overcome the heat and the cold

To cope with different temperatures, Koreans changed their clothes according to the season. Since the dawn of history, there seems to have been a gap between the clothing worn by the rich and the poor. No matter how cold it was, peasants wore thin cotton clothes quilted with cotton-wool. People with some money to spare could wear a vest made from animal fur, and those who could afford more wore leather garments.

Hot and humid weather, too, found the haves and have-nots dressed differently: the former in ramie; the latter in hemp. Both textiles were suitable to the extreme summer heat wave. On Jeju, though, a unique material was developed. Instead of hemp, the islanders created *garot*: cotton died with pounded unripe persimmons. When dried under the blazing sun in midsummer, the textile turned a reddish brown color and the clothes made with this cloth are called *garot*. Today this seems to have become a fashionable product favored by environmentally conscientious Koreans. But in the past, it was an essential element in the wardrobe of ordinary people on Jeju. No matter how much one sweats, *garot* does not stick to the skin, so it was perfect for the hot, humid Jeju summer.

When people speak of climate change, they say that the current climate

Jeju's unique material, persimmon died reddish brown cotton, is suitable to the extreme summer heat of the island.

is not as cold as in the past, and they emphasize that it was much colder in the 1960s and 1970s than it is now. But I can't help wondering if this is really true. Looking back, it certainly seems to have been colder back then, but I wonder how much of the difference is merely subjective, a matter of differences between current and earlier lifestyles.

Today, who would walk that as far to school as we used to? In cold weather, who would go out into the street in such light clothes? In the countryside, I walked a total of 6 km everyday to attend school. Today on that same road there are frequent buses; and cars come often enough that one could easily hitch-hike. Nobody walks except by choice. But in the past, there were no cars, only empty fields. Even the wealthy had no choice but to walk. The winter wind that swept across those fields would indeed have felt much colder than what we experience today.

The first time I wore aan overcoat of a famous brand, was in the mid 1980s. I was able to wear one with duck-down insulation only after I got married. But during those cold years in the past, no one, rich or poor, could wear duckdown coats. Such garments were not to be found in Korea. Those were times when people wore long underwear and a shirt, and a nylon jacket with a thin sponge layer over that. Wouldn't todays's weather also seem cold if you were walking through empty fields dressed so poorly? Today, even the clothes sold at street stalls look much warmer.

Today the gap of clothes between regions have narrowed to the point where it is nearly impossible to judge the wealth of an area by what the people are wearing. But when I first left Jeju, I worried a great deal about what I should wear when I went to Seoul. Having spent my childhood on an island where the temperature rarely dropped below freezing, I had many things to think about in preparing for my first journey to this unknown place called Seoul. Most important to me was what clothes I should wear. Weather casters at the time mentioned subfreezing temperatures on most winter days in Seoul, but I had no clue just how cold this 'subfreezing' would actually feel. I just vaguely thought, "That must be very cold." On Jeju, even when it was above freezing, it seemed plenty cold to me, so I was really worried about the cold in a place with temperatures below freezing every day.

With no better solution at hand, I simply put on the warmest clothes I had and boarded the ferry for the mainland. I took an express bus from Busan and arrived in Seoul late at night. It was interesting to me to see frost form on the bus window, and then freeze to thick ice. It was something I had never experienced back home. The bus window display seemed like a show put on to convince a rural boy that Seoul was indeed a cold place. As it happened, that was when a cold wave was gripping the whole world. It was hard to walk any distance. Although I was wearing several layers of clothing, my feet felt as if they were frozen, as I stood by the bus stop. My guide at the time made an effort to show me some of the sights of Seoul, but I pressed several times to just be taken to my lodgings.

But once I had finally arrived to the place where I would be staying, it was equally difficult to bear. The room was suffocatingly hot. As defense against the cold, the owners had covered all the cracks between walls and all the windows with sheets of vinyl. Moreover, the floor was so hot I could hardly sit on it, and there was a briquette-burning heater in the hallway. Never having encountered such a heating system before, I found it hard to breathe. On Jeju, where the cold was mostly bearable as long as you blocked the wind, the only mode of heating people used in their homes was to burn some firewood on a small brazier in the middle of the room.

Today, Koreans use boilers to heat their homes, whether in the city or the countryside. But not long ago, the *ondol* system of heated floors was the most common way of warming homes in winter. Unique to Korea, the *ondol* system involved smoke and heat from a fireplace in the kitchen passing through empty space under a floor made of a flat stone. The hot air from the fire heated the stone, and thereby the entire room.

The *ondol* was particularly well-suited to Korea, where the cold continental climate defines the winters, and the humid oceanic climate prevails in summer. The warm *ondol* floor that heated a room in winter also served to clear away the humidity in summer. Moreover, the smoke helped to drive annoying insects from the room.

Nowadays, Korean houses and apartments in the cities are well equipped to handle cold winters, with well insulated walls and double sliding windows. But if you travel into the countryside, you can easily find homes that remind you of the past, especially in mountain areas. In many cases, inhabitants spend the winter in the warmth of a house surrounded by piles of firewood. Just looking at those logs, a passing traveler can sense the warmth inside. The fragrance of burning firewood spreads pleasantly through the winter air in a mountain village, and you can see a fire burning

Farmers are gathered, making home-crafts of straw or bamboo, or playing a board game in a room made snug and cozy by the heated *ondol* in winter. (Yangyang, Gangwon-do. Jan., 2008)

in every kitchen of the quaint houses. If there are any children in the house, they will be roasting some sweet potatoes at that fire. And in the room made snug and cozy by the heated *ondol*, the villagers are gathered, making home-crafts of straw or bamboo. Such idyllic winter scenes, however, may be disappearing into the past.

Author

Seungho Lee is professor of Climatology, Department of Geography, Konkuk University, Seoul

Translator

Jinna Park received her M.A. in English Literature from Hankuk University of Foreign Studies. She currently lives in Seoul and works as a literary translator.

The climate and culture of Korea / author: Seungho Lee ;
translation: Jinna Park. -- Seoul : Purengil, 2010
   p. ;   cm

ISBN 978-89-6291-131-2 03980 : ₩ 30000

climate

453.9911-KDC5
551.69519-DDC21   CIP2010002138

The Climate and Culture of Korea

First Edition Jun 30, 2010

Author **Seungho Lee**
Translation **Jinna Park**

Publisher **Seon-gi Kim**
Published by **Purengil**
Registration Number **16-1292**
Registration Date **April 12, 1996**
3F Ujin B/D, 1001-9, Bangbae-dong, Seocho-gu, Seoul, Korea
E-mail **pur456@kornet.net**
Home-page **www.purungil.com**
Tel **+82-2-523-2009**
Fax **+82-2-523-2951**

Printed in Korea

ISBN 978-89-6291-131-2 03980

Korea Foundation
한국국제교류재단

The Korea Foundation has provided financial assistance for the
undertaking of this publication project.